The Relevance of
General Systems Theory

THE INTERNATIONAL LIBRARY OF
SYSTEMS THEORY AND PHILOSOPHY
Edited by Ervin Laszlo

THE SYSTEMS VIEW OF THE WORLD
*The Natural Philosophy of the New
Development in the Sciences*
by Ervin Laszlo

THE RELEVANCE OF GENERAL SYSTEMS THEORY
*Papers Presented to Ludwig von Bertalanffy
on His Seventieth Birthday*
Edited by Ervin Laszlo

HIERARCHY THEORY
The Challenge of Complex Systems
Edited by H. H. Pattee

Further volumes in preparation

THE RELEVANCE OF GENERAL SYSTEMS THEORY

Papers Presented to Ludwig von Bertalanffy on His Seventieth Birthday

Edited by ERVIN LASZLO

GEORGE BRAZILLER *New York*

For information address the publisher:
George Braziller, Inc.
One Park Avenue, New York, N.Y. 10016

Standard Book Number: 8076–0659–6, cloth
                       8076–0658–8, paper
Library of Congress Catalog Card Number: 72–81355

Second Printing
Printed in the United States of America

Preface

This book is based on the interdisciplinary symposium honoring the late Ludwig von Bertalanffy on his seventieth birthday, which took place on September 18, 1971, at the State University of New York College of Arts and Science at Geneseo.* The symposium was convened by the editor of the present book under the sponsorship of the Department of Philosophy and the Committees for the Geneseo Centennial Year.

The seventieth birthday of a great scientist and thinker is cause both for personal celebration and for a consideration of his major contributions. Those who have had the privilege to be personally acquainted with Von Bertalanffy during his long career in Europe, Canada, and the United States, extended to him greetings and congratulations on his anniversary. But the symposium itself was meant to reach beyond the personal sphere, and pay homage to the celebrant by examining the present state, and prospects of future development, of one area to which he has contributed perhaps more than anyone else: general systems theory. Although by this method Von Bertalanffy's other achievements—e.g., in the fields of cellular and comparative physiology of metabolism, animal growth, and cyto- and histochemistry—could not receive more than passing consideration, it was hoped that fresh light could be shed on developments in the fast-growing area of general systems research, which he founded together with Boulding, Gerard, and Rapoport. To this end, contributors were asked to review the

---

* Professor Von Bertalanffy died of a heart attack on June 12, 1972, in Buffalo, N.Y.

impact and relevance of general systems theory to their particular field of interest, and bring the fruits of their own insights to bear on problems and developments which could further profit from a consistent systems approach. Thus Anatol Rapoport reported on mathematics and systems thinking in general, Howard Pattee on problems of theory of evolution, Robert Rosen on methodological problems of systems description in biology, Ekkehard Zerbst on current work in physiology, Kenneth Boulding on economics, Lee Thayer on communications theory, William Gray and Nicholas Rizzo on psychology and psychiatry, and Lionel Livesey and Jere Clark on education. In the opening paper Ervin Laszlo surveyed the parameters of the contemporary systems movement, and in his response Ludwig von Bertalanffy offered an overview of the materials presented and entered a much needed plea for humanism.

The resulting book is an important contribution to the recently launched INTERNATIONAL LIBRARY OF SYSTEMS THEORY AND PHILOSOPHY. It is testimony to the diversity of applications of general systems theory, its unity of conceptual structure, and its vast and in part still unexplored potentials. It is dedicated to contributing to the erosion of disciplinary encapsulation, and to pointing the way toward further interdisciplinary research and humanistic application in the many areas of science and philosophy where systems concepts are relevant and the need for holistic and yet rigorous thinking pronounced.

E. L.

Geneseo, N. Y.

Contents

INTRODUCTION

The Origins of General Systems Theory in the Work of Von Bertalanffy

ERVIN LASZLO

SOME forty-five years ago, in his middle twenties, Ludwig von Bertalanffy formulated certain basic concepts concerning the organism as an open system. These concepts, in recent years, have grown into a full-scale "revolution" in many sectors of scientific thought. In 1928, two years after receiving his doctorate at the University of Vienna, he published *Kritische Theorie der Formbildung* (in English translation, *Modern Theories of Development*). In this first book Von Bertalanffy pointed out that "the chief task of biology must be to discover the laws of biological systems (at all levels of organization)." And he added, "We believe that the attempts to find a foundation for theoretical biology point at a fundamental change in the world picture." At that time his main concern was with the life sciences, more specifically with the advent of "organismic biology" and "the system theory of the organism." In subsequent writings he elaborated such a biological theory in some detail. Here his two well-known books must be mentioned, *Theoretische Biologie* and *Problems of Life*. In his mature years, he extended his organismic open system theory into psychology (in *Robots, Men, and Minds* and in *Organismic Psychology and Systems Theory*) and, in a summation of the many-sided relevance of his conceptions, provided a "bible" for the rapidly blossoming systems movement (in *General Systems Theory: Foundations, Development, Applications*).[1]

Von Bertalanffy, like many men of genius, was never content to explore one avenue of thought or one area of research, however broad, to the exclusion of others, and his interest and contributions ranged from his famed equations for animal growth, widely used in fisheries the world over, through fluorescence cytodiagnosis for the early detection of cancer, to the origin of the postal service in fifteenth-century Italy. Within that range, he managed to deal with, and make important contri-

3

butions to, philosophy—especially theory of value, symbol, and culture—social and behavioral science, and psychology. His own thought pioneered the burgeoning lines of development of the movement which he initiated, as he explored now one of its branches, now another.

The concepts Von Bertalanffy offered in his twenties were consolidated by him first into a biological theory, then a general theory with interdisciplinary applications ("General Systems Theory"). Thanks to his work, and to the genius of others who have joined forces with him (Boulding, Rapoport, et al.), we have today a rapidly growing school of thought, variously referred to as the "systems movement," the "systems approach," or "systems research." This complex of theories, concepts, and commitments, signifies, as Von Bertalanffy surmised in 1928, a change in the world picture. Or, as he noted in 1967, it constitutes a "new natural philosophy." I now prefer to speak of it as "a new paradigm of contemporary thought."[2] This concept is justified by Kuhn's explication of what is meant by a paradigm. A paradigm, Kuhn points out, is a concept that denotes something far wider in nature and scope than a theory. It is, rather, a "disciplinary matrix": "disciplinary" because it refers to the common possession of the practitioners of a particular discipline; and "matrix" because it is composed of ordered elements of various sorts, each requiring further specification.[3]

The Disciplinary Matrix of General Systems Theory

Let me consider briefly the "disciplinary matrix" which is shared by systems theorists and which defines their adherence to the systems movement originated and championed by Von Bertalanffy. First of all, it is necessary to emphasize that the discipline in question is not a conventional kind. Conventional disciplines treat one particular subject matter—a well-defined segment of the range of empirical phenomena, or of culture, or symbols. The systems movement, however, constitutes a discipline in virtue of the agreement of its practitioners that their

4

problems and investigations cut across narrowly defined tra-ditional disciplinary boundaries. The systems movement is an "interdisciplinary discipline"; a specialty traversing traditional specialties, and an open avenue toward general theory, linking and integrating the fragmented pieces of contemporary scien-tific thought.

Secondly, "matrix," in "the disciplinary matrix of the systems movement," refers to a set of shared presuppositions in the form of principles and conceptual approaches, to which the practitioners of this movement manifest a deep-seated commit-ment. I find the following components to be among the most basic:

1. *Holism as a methodology, and even an ontology.* Since the time of Galileo and Newton, modern science has been dominated by the ideal of explanation by reduction to the smallest isolable component's behavior in causal terms. Phe-nomena, however complex, were sought to yield isolated causal relations, and the sum of these were believed to constitute an explanation of the phenomena themselves. Thus two-vari-able linear causal interaction emerged as the principal mode of scientific explanation, applying to the primitive components of a given complex of events. Explanation in these terms pre-supposed atomism and mechanism as a general world view. But when contemporary science progressed to the rigorous ob-servation, experimental testing, and interpretation of what Warren Weaver called "phenomena of organized complexity," such explanations no longer functioned.

Complex phenomena proved to be more than the simple sum of the properties of isolated causal chains, or of the properties of their components taken separately. Such phenomena, as Von Bertalanffy pointed out, must be explained not only in terms of their components, but also in regard to the entire set of relations between the components. Since such relations can-not be treated with mathematical rigor with the classical con-cepts (which soon encounter their limitations in the "three-body problem"[4]), new concepts and approaches were sought. Characteristic of these is the assumption that sets of related events can be treated collectively, as systems manifesting functions and properties on the specific level of the whole. The holistic approach proved to be enormously fruitful in biology, where it was first explored by Von Bertalanffy, and found applications of great promise likewise in the social and behavioral sciences. In consequence an increasing number of investigators adopted holism as a methodology. Those among them who were philosophically oriented saw in holism not merely a competent methodological instrument, but also a valid conception of the empirical world. This conception now functions as a basic presupposition in the field of contemporary empirical systems research.

2. *Integration of scientific knowledge as an ideal with real possibilities of realization.* The potential of systems theory in this regard is the applicability of holistic systems descriptions in different realms of investigation. The emerging isomorphies in special scientific theories can be conceptualized as invariances within a general systems meta-theory, thereby relating and integrating hitherto diverse concepts and theories.

3. *Unity of nature as a philosophical credo.* The observed isomorphies in theory lead, in any but the most skeptical viewpoint, to the assumption that the empirical phenomena described and interpreted in the theories manifest analogous strands of order. Hence basic reference points can be indexed for a natural philosophy which perceives the world as governed by the same kind of fundamental laws and principles in all its different realms. Such a view does away with the traditional dualisms of mind and body, man and nature, and individual and society.

4. *Humanism as a task and responsibility of science.* A growing consensus of feeling among systems theorists today concerns our ability to come up with reasoned and reliable solutions to the problems which face our species. Many see the answer in the possibility of a "design theory"—of education, the social order, the economy, or the ecology—based on models

6

of man, nature and society. The wide scope of this movement, which encompasses a systems *philosophy* at its qualitative-general tip, and a systems *technology* at its quantitative-specific base, permits cautious optimism in the assessment of the chances that operational programs could be developed for application to concrete societal problems, translating the descriptive and normative content of a general system theory into practical terms. Certainly the ability of this movement to transcend traditional boundaries makes it uniquely qualified to face this great challenge and grave responsibility.

The disciplinary matrix of the systems movement can be specified in much more detail and with much greater penetration (see, for example, the works already mentioned of Von Bertalanffy and Laszlo). The above key concepts suffice, however, to characterize the nature of its paradigm in reference to shared commitments to basic presuppositions.

The Scope of General Systems Theory

Contemporary systems research cuts across the following boundaries, hitherto largely unpenetrated in contemporary systematic thought: (1) traditional disciplinary limits; (2) cultural-ideological barriers; (3) quantitative-qualitative, or "science-philosophy" gaps; and (4) descriptive-normative, or "pure vs. humanistic science" distinctions.

1. Systems research is undertaken by men belonging to a wide range of traditional disciplines. Indeed, when this writer surveyed the home affiliations of the participants at recent conferences in New England, California, and the Soviet Union, he found that few shared a common disciplinary background. Such is precisely the case in the present symposium. A similar finding comes to light if one surveys systems theoretical publications, such as the *General Systems Yearbooks.* This is an entirely normal state of affairs, since the movement is, by its very nature, interdisciplinary. Its practitioners come from all branches of the natural and social sciences, pure and applied, as well as from the humanities.

2. Systems work is carried out today in formerly distinct, and even opposed, cultural-ideological climates. At the Thirteenth World Congress of History of Science in Moscow (August 18–24, 1971), papers presented by systems theorists showed no differences in content directly ascribable to ideological-cultural factors. Rather, there appeared a striking parallelism in work done in the United States and the Soviet Union, and something of a gap as regards contributions from western Europe, especially the Continent. However, visits in recent months by this writer to West Germany and Sweden showed that interest in these countries for systems theory is pronounced.

This is not to suggest that cultural-ideological influences are null and void in regard to the systems movement; rather, that they do not conflict with this movement's disciplinary matrix and hence do not stunt its evolution. Its particular developmental pathways do manifest traces of particular cultural-ideological constellations. For example, in Poland a "science of science" aspect is stressed with strong logical-formal elements, testifying to the influences of the school of famed Polish logicians and their relative freedom to pursue their own theories. In Hungary, on the other hand, a "systems aesthetics" school is taking shape, recruited from art-oriented Lukács followers and systems-oriented social and behavioral scientists, engineers, and younger philosophers. There is a science-methodological emphasis within the Soviet systems school itself, which is due to the fact that academic philosophers in the U.S.S.R. have tighter ideological commitments than systems theorists, who work mostly in institutes of science and history of science.

3. The systems movement extends over several conceptually distinct levels, which interact in virtue of their shared disciplinary matrix. Following Von Bertalanffy's assessment of trends in general systems theory,[5] we can distinguish two levels on the quantitative-qualitative or science-philosophy scale. On the quantitative level of systems *science,* work progresses both in

"pure" and "applied" theory. On the formal end of the spectrum, pure systems science encompasses mathematical systems theory. Work here specifies systems concepts and systemic relationships, developing the tools and principles for assessing systems phenomena in rigorous logical and quantitative fashion. Eminent representatives of this fast-growing field include Anatol Rapoport and Robert Rosen. In the empirical systems science area, research constitutes a systematic effort to state problems encountered in a given field of science in systems terms. Hereby both the problem-solving capacity of the given science is enlarged—since systems formulations are often capable of resolving issues and explaining observations which remain anomalous for conventional theory—and the concepts and theories of the field become extensible to, and readily comparable with, concepts and theories in other systems sciences. Outstanding examples of empirical systems science research are the works of Kenneth Boulding in economics, Zerbst in physiology, Livesey and Clark in education, Thayer in communication theory, Gray and Rizzo in psychology, and Pattee in biology.[6]

Quantitative systems science comprises, in addition to pure theory, applied theory, or systems *technology*. Under the various headings of systems engineering, management science, operations research, computer science, etc., systems concepts are translated into operational terms. This includes developing the software (operational models in different fields) and the hardware (control technology and computerizing) for confronting concrete problems with practical solutions.

Systems science, pure as well as applied, seeks specific answers to particular problems with the greatest possible precision. Hence it is quantitative in orientation and usually mathematical in form. Systems *philosophy*, on the other hand, while likewise seeking the greatest possible clarity in its formulations, is of necessity on the opposite side of the quantitative-qualitative, specificity-generality spectrum. Systems philosophy is, typically, an integrative "linkage" discipline. It seeks the interconnections of independently researched and formulated theories; it probes the basic texture and ultimate implications of the systems paradigm. By providing heuristic hypotheses for more specific scientific theory formulation, it guides the imagination of systems scientists and provides an explicit general world-view the like of which, in the history of science, has proven to be most significant for asking the right questions and perceiving the relevant states of affairs.

Systems philosophy encompasses traditional philosophical fields, such as ontology, ethics, natural philosophy, and epistemology, and reassesses the traditional concepts in reference to newly emerging systems conceptions. It is practiced, for the time being, more by scientists with wide perspectives than by academic philosophers. (Compare the writings of Von Bertalanffy, Waddington, Bohm, Gerard, Blauberg, and Miller on the one hand, and Laszlo on the other.) In Communist countries the reason for this lies in the strong ideological commitment of academic philosophy, while in the Western world it is a classic case of information lag in the channels of communication between systems science and the philosophical community.

4. The contemporary systems movement overcomes the time-honored gulf between descriptive and normative theory and links the realms of fact and value. Systems theorists investigate dynamic, goal-oriented systems, with definite if flexible programs for coping with their environment and assuring their growth and development. Thanks to Wiener's pioneering work on the concept of "purpose," it is now possible to *describe* systems in terms of ends and goals. Such descriptions can be, and in fact are, offered by investigators of systems on many different levels. (For example, the human organism is a goal-guided system on one level; its cells and organs are that on another, and a university is such a system on still another level.) Out of an understanding of the goals and ends of systems (as the conditions which their programs and "preferences" point toward) and of their interactions within environments shared with and provided for one another, grows a general

10

comprehension of the role of human values in concrete circumstances. An integrated systems approach can not only tell us that values are objective factors in behavior and development, but can show which values lead to the actualization of human potentials, and which bring about conditions that frustrate and thwart the individual. Hence from a *descriptive* science of the role of values in guiding the behavior of men in nature and society, we get anchor points for a science of *normative* values, or value-norms. In this sphere of systems theory there is a meeting of pure and of humanistic science, as the former becomes the ground for the latter.

I conclude that the systems movement, initiated and vitally furthered by Ludwig von Bertalanffy, has come to represent a new paradigm of contemporary scientific thought. It has unity through diversity, as the papers assembled in this volume clearly demonstrate; it is an interdisciplinary discipline, and an integrator of diverse fields of knowledge in diverse cultural and geographical locations. It is pursued on different levels, but with a remarkable consensus of basic commitments. And it provides possibly the most powerful tool we have today for effecting the unification of scientific knowledge, and the utilization of that knowledge for humanistic ends.

1

The Search for Simplicity

ANATOL RAPOPORT

SIMPLICITY is an aspect of structure. To be perceived, however, structural simplicity must "mesh," so to say, with a cognitive act. The result is recognition, experienced subjectively as a catharsis (a release of tension). It seems that the inquiring mind actively seeks such "meshings," as if driven by a libido. The delight of unraveling a puzzle, of guessing a riddle, of solving a problem, and of perceiving regularity where none was apparent before, are all of one kind.

Literacy in mathematics, even on the most elementary level, can be a rich source of experiences of this sort. Consider the sequence

$$5/8, 2, 3, 9, 6, 15, 9, 21, 39, 15, \ldots$$

Trying to make sense of this sequence can be a frustrating experience. The numbers seem to be neither steadily increasing nor decreasing. Most, but not all, are odd. The 39 seems "less fitting" than any of the others, except the 5/8, which looks altogether out of place. Nor does it help to learn that the next number of the sequence is 51.

Yet the sequence has a simple structure, embodied in the formula $\frac{1}{8} (p_n^2 - p_{n-1}^2)$, which generates the sequence, the p_n being the successive primes. Once this is recognized, everything suddenly becomes clear. It is now clear why all the numbers except the first are whole numbers. This is because every prime, except the first (2) is odd. Consequently the differences of the successive primes (after the first difference) must be even, and, moreover, the differences of the squares must be divisible by eight. This is so since

$$p_n^2 - p_{n-1}^2 = (p_n - p_{n-1}) (p_n + p_{n-1})$$

15

so that the left side is divisible by 8 if at least one of the factors is divisible by 4 (since both factors are even). Suppose $p_n - p_{n-1}$ is *not* divisible by 4. Then, being even, it must be of the form $4m + 2$, where m is an integer. But then the second factor is $4m + 2 + 2p_{n-1} = 4m + 2 (1 + p_{n-1})$ and divisible by 4, since $1 + p_{n-1}$ is even, so that $2 (1 + p_{n-1})$ is divisible by 4. Similarly, if the second factor is not divisible by 4, the first must be.

If this explanation gives you pleasure, the pleasure would have been more intense if you had found it yourself. Then you would have experienced the catharsis, the so-called "Aha!" phenomenon, the crowning reward of a successful search for simplicity.

It is, of course, not necessary to love mathematics to experience very similar pleasures. All cultures have proverbs and riddles. The proverb illustrates a general principle by a special case. A riddle is an invitation to translate a metaphor and by doing so to perceive the similarity of apparently dissimilar things. Both are exercises in the search for simplicity.

"Too many cooks spoil the broth" and "With seven nurses, a child loses an eye" say exactly the same thing, namely, that in many situations a single decision-maker is preferable to a committee. But to say it in this prosaic way is to fail to take advantage of an important pedagogical principle, that of inviting *the other* to make the generalization himself and so to experience the insight. The pleasure he experiences will make him more receptive to the idea expressed in the proverb. Jesus knew this principle. That is why he resorted to parables.

The riddle, too, invites a discovery. "What has a face but no mouth, hands but no arms, can run without legs?" The image of the clock calls attention to the metaphorical aspect of even the most commonplace words. The insight is the reward for guessing the riddle.

The heart and soul of poetry are, of course, metaphor. The poet thus calls attention to the similarity of widely dissimilar things or events. The quality of the poetic metaphor resides in its originality and its aptness. Originality ensures that the insight afforded by the metaphor is fresh; the aptness insures the acceptance of the insight, a feeling that "it is indeed so."

Insight, then, is a sort of fusion of inevitability and surprise. It has been said that Mozart's music has that quality. In hearing it, one feels that "it could not be otherwise," but only in retrospect. Classical art is based mainly on the aesthetic principle of *discovered* simplicity.

The solved problem, the guessed riddle, the perceived similarity evoked by the metaphor are all "obvious" *after* the insight. The general feeling, then, is that the world is simpler than it first appeared to be. The search for simplicity is the search for such insights.

It is easy to relate this search to the pleasure principle, but such an explanation, being teleological, perhaps tautological, does not go to the heart of the matter. By and large, activities that provide pleasure independently of cultural setting are survival-enhancing activities, eating and sex being the most obvious examples. With regard to man's enormous variety of pleasure-giving activities, it is sometimes not so easy to decide whether they are independent of cultural setting and, therefore, probably related to survival. In fact, many pleasure-giving or, at least, pleasure-seeking activities, as we know so well, jeopardize rather than enhance survival. We can, perhaps, assume that these are specific cultural distortions of activities that stem from survival-enhancing drives. A strong case can be made for the search for simplicity as an activity rooted in a survival mechanism. A simple, predictable environment is easier to adapt to than a complex, capricious one. For most animals, "learning to predict" is a matter of passive conditioning. In some, however, we see evidence of curiosity. Rats and cats explore unfamiliar surroundings. Dogs don't just react to smells as they come; they go sniffing around. Primates *examine* objects. All these activities seem to be directed toward making the unfamiliar familiar, hence rendering a complex environment simpler.

However, whether other animals experience the quasi-orgasm of the insight the way we do, we cannot know. So far, our knowledge of nonhuman psychology is confined to what we observe of animal behavior. Projection of our introspective impressions on nonhumans is not reliable. Nevertheless the evidence of curiosity, the active acquisition of "knowledge" in other animals (at least, in the primitive sense of making the unfamiliar familiar), leads to the reasonable conjecture that the characteristically human search for simplicity is based on a biologically determined drive.

Science is clearly a systematized search for simplicity, a method of making the world predictable. The most conspicuous feature of science, which, in popular estimation, both explains its origins and justifies its existence, is that it represents the harnessing of the search for simplicity to other purposes, specifically to gain power over the environment. Understanding the world and controlling it are logically separable. For instance, when we have understood the structure of the number sequence above, we have not thereby gained control over anything. Understanding the motions of the planets does not confer the power to control them. Nevertheless, there is an undeniable connection between understanding and control. Understanding the nature of the world can confer power over a portion of it.

But, aside from these utilitarian aspects of understanding, there is still another connection between understanding and control. Knowledge of the world involves knowledge of causes and effects. Such knowledge is accumulated much more rapidly and systematically and also becomes more reliable if the presumed causes can be manipulated while presumably irrelevant events are excluded, as in controlled experiments. Therefore, control of portions of the world becomes desirable not only for the purpose of exploiting the environment but also for the purposes of understanding it.

Such is the search for simplicity as it is embodied in science. Control has become incorporated into it to such an extent that it has become an integral part of understanding, as "understanding" is understood in science. But the desire for understanding need not be welded to the desire for control. The two have been linked in a means-ends relation, but it is characteristic of human strivings that not only can ends become means to other ends but also means can become ends in themselves. In fact, curiosity has probably antedated rapacity (the obsession with power) in the development of the human psyche, since familiarity confers the survival-enhancing ability to predict, independently of the ability to control.

With regard to the relation between understanding and prediction, rather perplexing questions arise. First, is prediction the crucial criterion of understanding? In the positivist philosophy of science, which purports to be also a philosophy of knowledge, the answer is given in the affirmative. In this philosophy, reality is pictured as the totality of *events*. Consequently knowledge of reality is knowledge of events; more precisely, of the relations among events. Accordingly, all meaningful assertions about reality are either tautological, i.e., equivalent to definitions, or predictive in the form "If so, then so." The test of understanding a portion of the world is a test of the ability to predict something on the basis of the alleged understanding.

In this view, the internal, phenomenological aspect of understanding is ignored, and for good reason. Access to the internal psychological states of others is denied, and introspection is not a reliable source of knowledge. Consequently "objective" knowledge must be publicly, not privately, verifiable.

Yet there is resistance to this epistemological view even in circles claiming a niche in the scientific enterprise. I am referring to those conceptions of social science where prediction is declared to be not a necessary criterion of understanding and where on occasions the very possibility of predicting human events on a significant scale is denied.

As an example, take the traditional view of the method of history. The historian's unquestioned task is to reconstruct events of the past on the basis of available records. Most historians, however, will also agree that the "understanding" of historical events is also part of the historian's task. Many historians dissociate this sort of understanding from the ability to predict anything. And yet, causal explanations are accepted by the same historians. That is to say, the historian feels that he has understood a historical event if it appears as an *expected* consequence of a configuration of other events. So far, this interpretation of understanding does not differ from the scientist's. The historian, however, maintains that, even though he understands the events, that is, the causal connections between an event and those preceding it, he cannot utilize this understanding to predict events, because the configuration of the events in question was unique; it was the first such configuration in history and is not likely to be repeated. Hence the "If so, then so" paradigm of science is vacuous in historical settings. The "so" of the antecedent may never recur.

How, then, does the historian see the event explained as an expected consequence of the configuration preceding it? (For that is what it means to explain an event.) If the configuration was unique, he could not base his expectations on previous experience. The conclusion is inescapable that the historian's explanation is based on an analogy. There must have been something in his experience analogous to the allegedly unique configuration of events. The configuration is called unique because its analogues are not involved in the explanation. The historian may not even be conscious of evoking such analogies. But he must be evoking them, or else how does he know, at least in retrospect, that the event was an expected consequence of its antecedents?

To take an extreme case, consider the historian who refuses to recognize any sort of historical determinism, who believes that the course of history at any given moment is determined only by purely accidental configurations. He is the historian who takes seriously the nursery jingle, "For want of a nail, a shoe was lost; for want of a shoe, a horse was lost . . . ," and so on with the rider, the message, the battle, and the kingdom. Even this historian invokes causal principles; for example, that horseshoes not firmly fixed with the requisite number of nails often fall off; that a horse without a horseshoe is likely to stumble and fall; that the outcome of a battle depends on the information (in this case, the message borne by the rider) available to the commander; that outcomes of battles sometimes have serious political consequences. All these are respectable bits of knowledge, and they are all obtained from experience. What the antitheoretical historian refuses to do is to seek causal connections between *large* recurring events of the sort that can be fitted into a *theory* of history. He nevertheless depends on analogies for his explanations, such as that one horseshoe is like another and even that one kingdom is like another. Else why does it appear natural to him that the kingdom was lost as a result of a battle? Either he knows of more than one kingdom that was so lost or he draws on an unconscious analogy between a kingdom and something else, for instance, a city, several having been lost following lost battles.

In short, all understanding stems from perceived analogies—recognition that something is like something else. The positivist scientist's understanding is not excluded. For in the most explicit predictions of the form "If so, then so," the many instances of the antecedent are assumed to constitute the same initial condition. In the statement "If a spark passes through a mixture of hydrogen and oxygen, the gases will combine, forming water and releasing heat" it is assumed that oxygen is oxygen, hydrogen is hydrogen, a spark is a spark, and so on.

I venture to say that the most fundamental difference between the physical and the social sciences is that, in the former, reliable analogies are comparatively easier to discover than in the latter. In the biological sciences, the situation is intermediate. Since the physical sciences deal with matter and events involving bits of matter, analogies among physical events are sometimes so obvious that they appear to be identities

21

rather than analogies. One falling stone is very much like another. Substances are combinations of scores of elements, consisting of particles that can be assumed to be identical. Still, even in physics, it took a stroke of genius to discover that the motion of the moon and that of the falling apple are analogous; that burning wood and rusting iron are analogous processes, and so are the propagation of light and of electromagnetic waves.

Biological analogies are deeper and come to light later. Like the physical analogies, they simplify the world picture. However, the sweeping generalizations of evolutionary theory did not confer on man the power conferred by the physical sciences. The predictions of biological theory are not as neat and exact as those of physical theories, and for good reasons. The predictions of physical theories, for the most part, concern situations where initial conditions can be precisely specified. If such initial conditions are not found in nature, they can be arranged. Such arrangements are considerably easier to realize with inanimate than with animate matter, because the properties of animate matter are much more sensitive to being tampered with than those of inanimate matter. In particular, living tissue in vitro may behave quite differently than in situ. Controlled biological experiments are, of course, possible, but they are more difficult and their scope is more limited than that of physical experiments. For this reason, biology has had to depend to a greater extent than physics on theories of larger speculative scope, in which reasoning by imaginative analogy plays a more important role.

In the social sciences, this role must be even greater. This is not to say that in the social sciences controlled experiments cannot be performed. They can and are, but their relevance to the large, important issues of social science becomes the more questionable, the "cleaner," more tractable, and the more rigorously controlled the experiments become.

There are two ways out of this bind. One is taken by the positivist social scientists who confine their attention to clearly objective, quantifiable, manipulable data, eschewing whatever concepts cannot be related to such data. Thereby they lay themselves open to accusations of trivializing social science. The other way out is to dissociate "understanding" from predictions and control and to declare that the social sciences are based on epistemological principles different from those of the natural sciences. The proponents of the *Verstehen* school of social science stand in danger of succumbing to the sterile methods of the system-building metaphysicians, who peddled their fantasies as revelations of profound truths about the nature of reality and of the universe.

It is in the light of this dialectic opposition between exact knowledge and speculation, between the specifically demonstrable and the intuitively perceived that I want to examine the import of general systems theory to which Ludwig von Bertalanffy has made outstanding contributions.

To repeat, in the physical science, the trade-off between simplicity and generality does not apply. On the contrary, to the extent that the physical sciences become mathematicized, simplicity not only does not disappear in generalizations but actually results from them, thanks to the unique property of mathematical language. This is seen most clearly in pure mathematics.

The simple formula $\frac{1}{8}(p_n^2 - p_{n-1}^2)$ embodies in it the at first sight unmanageable complexity of the number sequence generated by it.

The generalization of the concept of number to include imaginaries simplified the theory of functions, for instance, by revealing the analogies between exponential, hyperbolic, and trigonometric functions and the connections between the latter and the doubly periodic elliptic functions.

In the same way, the formula

$$F = K \frac{m_1 m_2}{r^2}$$

embodies in it the entire complexity of celestial mechanics. The triumph of mathematical physics was achieved by the simplifications through generalization.

However, as Kenneth Boulding once remarked, nothing fails like success. The attempts to transplant the method of generalization, brilliantly successful in the physical sciences, into biological and social sciences led to many disappointments, at times to delusions engendered by wishful thinking. I submit we must face the fact that there are no biological or social "laws" that are direct analogues of the laws of motion, the law of gravity, the conservation laws of energy and mass, the law of increase of entropy in isolated systems, etc. At most, there are *models* of specific biological and social phenomena, expressible as mathematical formulae to serve as working hypotheses. The models bring some simplification into the study of phenomena of which such models are reasonable approximations. If it happens that several phenomena of widely different content can be described by the same mathematical model, both simplification and generalization have been effected. Therein lies the power of *mathematical* general systems theory, based on the principles of mathematical isomorphisms, the completely rigorous version of analogy.

Since my own predilections are toward mathematicizing, I have usually emphasized the role of mathematical systems theory in the epistemology of science. However, I should not like to see the other epistemological contribution of general systems theory neglected, namely, the organismic concept of system. In contrast to the mathematical concept, which defines a system as a set of relations among variables that are defined or postulated, the organismic concept depends on an act of intuitive recognition. In the most conspicuous instance, we experience a direct, intuitive recognition of a biological organism as an entity. This ability is evident even in higher animals. Of course, this ability does not depend on any conscious selection of variables and of relations among them: it is simply given to us, as it is to other animals.

24

The question now is, how far can this recognition be stretched? What else besides biological organisms can we get to recognize as "systems"? And how do we go about recognizing "theoretically fruitful" analogies in the structure, behavior, and history of portions of the world—material or ideational—that deserve to be called "systems"? And what is a "theoretically fruitful" analogy anyway? For instance to me it seems that the medieval comparison of the body politic to the human body (king = head; army = arms; peasantry = back; church = heart, etc.) was naive and sterile. But a more general comparison of a social system to a biological organism (drawing analogies between the respective functions of maintenance, defense, homeostasis, communications, and the like) is fertile and enlightening. Am I deluding myself, as the medieval philosophers did, or are the apparent analogies between social and biological organisms rooted in reality?

Again, it appears to me that the hypertrophy of military establishments is analogous to malignancies in living organisms. I think of malignant growths as results of an uncoupling of certain tissue cells from the regulating mechanisms of the organism. In consequence, the malignant cells become parasitic on the organism. As a rule, parasites introduced from the outside can be combated by the immunological reactions of the host against the biochemically "foreign" invaders. But when the parasites are the cells of the organism itself, the immunological mechanism fails. The malignant cells are recognized as "one's own." In general, *successful* parasitism depends on a simulation by the parasite of characteristics that the host "recognizes" (mistakenly) as its own. Consider the parasitism of the European cuckoo. It works because the cuckoo's eggs are mistaken by the hosts as their own.

Recently I saw a report of behavioral parasitism observed among ants. An insect "learned" to simulate the stroking actions that ants use on other ants in order to be fed. Can it be that the military establishments of the superpowers, by simulating a "protective function" which in the past may have been

25

genuine, have carved out a parasitic niche for themselves and an opportunity for unimpeded growth to the detriment of the host? And can it be that the reason these parasites cannot be "identified" as such and eliminated is because the "malignant cells" are the society's own rather than clearly foreign bodies? Is there some element of reality in this conception or is it simply dictated by my disdain for what the military represents in this era of supertechnology and of a shrunken globe?

To take an example removed from strong moral commitments, how valid and how far-reaching is the analogy between linguistic and biological evolution? Observe how closely the taxonomy of languages parallels the taxonomy of biological groups. Dialects are subspecies, languages species, grouped into families of various degrees of inclusiveness. Next, the evolutionary bifurcation of languages is conclusively demonstrable by historical evidence. The question is, how far can this analogy be pursued? Is there an analogue of natural selection in linguistic evolution? At first thought, it would seem that looking for an analogue of natural selection is going too far. Natural selection operates on phenotypic variations in the progeny of an organism. If a language is an "organism," what is its progeny? Certainly not the languages evolving from it. They are far too few to constitute the numerically vast generations on which natural selection operates. It takes hundreds or thousands of generations for natural selection to produce a new species, not to speak of larger taxonomic units. Surely no such plethora of generations is observable in the "progeny" of a language.

Nevertheless, the analogy between linguistic and biological evolution is so striking that it is difficult to stop looking for further dynamic connections besides the static structural ones. Such connections might be established if one took as the analogue of an organism the single utterance of a recurring linguistic unit, say, a word. This "organism" has only an ephemeral existence—the moment when it is spoken—but it has a progeny, namely, its subsequent utterances by the same or other speakers. For a word continues to "live" as long as it is repeatedly spoken. The frequency with which it is spoken represents the numerosity of its progeny. Moreover, the "progeny" exhibits phenotypic variations, slight changes in pronunciation or, perhaps, in the denotative or the connotative meanings of the word. *Now* natural selection can be seen to be operating. The differences that do not fit in the established patterns of speech (the "environment" of the word) will be eliminated by not being repeated. Occasionally, however, a "favorable" mutation may occur, that is, one that better fits the word into the newly emergent patterns of pronunciation, denotation, or connotation. The mutant will have progeny, may be "selected for," and may eventually establish itself as the successful variant. The totality of such changes is the evolution of the language as a whole.

Here, then, is a theoretical idea based on the exploitation of a suggested analogy. The validity of the idea is difficult to evaluate. I cannot think of a controlled experiment that would corroborate or refute the idea. Nor is it amenable to formulation in terms of a mathematical model, from which testable, quantifiable predictions can be drawn. But it is not an isolated idea. It applies equally to the evolution of artifacts. The progeny of an artifact is its copies. Copies contain variations, departures from the original design. The "environment" of the artifact is the human culture in which it is produced. A variant is better or worse adapted to this environment, depending upon the interaction of the artifact with man. The more adapted variants will be "selected for" and thus will have progeny (will be reproduced). Maladapted lines of descent will become extinct. The lighter-than-air flying machine, for example, is virtually extinct, and the giant types, like the mammoth dinosaurs, have disappeared from the face of the earth. Only small blimps, used as advertising billboards, continue a humble existence, like the lizards, distant cousins of the once great dinosaurs.

So pervasive is the analogy between the evolution of artifacts and that of organisms that even vestigial parts are noted in the former. The first horseless carriages were equipped with non-functional whip slots. The useless buttons on sleeves of modern jackets were once functional. Before the running-board completely disappeared from cars, a mere suggestion of it still persisted for some "generations."

The quest for simplicity stems from a conviction that underlying apparent wide dissimilarities are profound similarities, which, when one perceives them, make order out of chaos, hence simplicity out of complexity. More widely dissimilar entities than a living organism and a spoken word are hard to imagine. Yet, when both are considered as elements in large systems—a species in the one case, a language in the other—their analogous positions in the scheme of things become apparent. The scheme of things appears simpler than before.

General systems theory, in generalizing the notion of system, has instigated speculations of this sort. To the extent that the speculations are not yet amenable to validation in the scientific sense of the word, this aspect of general systems theory should not, I suppose, be subsumed under "theory." If a theory is a collection of theorems, as it is in any exact science, then general systems theory is not a theory. I cannot think of a single thing to say that applies to all systems, except tautologically. That is, properties of systems that apply to all systems are those that are consequences of the *definition* of a system, hence devoid of empirical content.

In principle, empirically verifiable assertions can be made about *a given class of systems*. If so, the task of general systems theory appears to be that of creating theoretically fruitful taxonomies of systems. Some theoretically important classes have indeed been singled out. Credit goes to Ludwig von Bertalanffy for emphasizing the very fundamental distinction between open and closed systems and the way in which many properties, once thought to be exclusive characteristics of living systems, turned out to be properties of open systems. In this way, continuity was established between the living and the nonliving. Workers in the field of nonequilibrium thermodynamics singled out an

28

important property of a class of open systems, those that satisfied the Onsager equations, and derived for these a generalization of the Second Law of Thermodynamics, according to which the rate of entropy production of the system is minimized in the steady state.

In these investigations, the physical properties of the systems are still central to the theory. However, the investigations can serve as a point of departure for a systems theory in which the *structure* of systems is taken into account. Such, for example, are the theories of electrical networks. Here again, the system point of view serves as leverage in suggesting intriguing questions concerning systems governed by other than physical laws. How does one classify organizations in terms of their structure? Do structural properties impose certain patterns of behavior and development?

Some years ago it was noted that the components of an organization concerned with its external relations tended to be proportional in size to the ⅔ power of the size of the components that have to do with the internal relations. If organizations were geometrically similar three-dimensional bodies, the areas of their surfaces would be proportional to the ⅔ power of their volumes. Since the surface is the interface of a body with the outside, and the volume is a measure of the inside, the relationship suggests an analogy. In fact, Bertalanffy's equations of growth, derived from the principles of anabolism and catabolism, are based on the same idea. Is the ⅔ power "law" of organizations a coincidence, an artifact, or a reflection of an analogy between an organization and a three-dimensional body? If so, how are the dimensions to be interpreted?

In pursuing investigations of this sort, it is well to keep in mind that most of them will lead to disappointments. We do not really have any serious "system laws" on which to build a grandiose theoretical edifice comparable to the edifice of mathematical physics. The search for simplicity, however, is seductive. It is easy to delude oneself into thinking that one has discovered a great universal law, and delusions of grandeur of this sort are—alas! too frequently—apparent in the work of scientifically or mathematically semiliterate cranks, whose herculean but fruitless efforts are the modern versions of the search for the philosopher's stone. As is the case with all passions that direct human energies, the line between creative and destructive, or self-defeating, effort is thin. The craving for knowledge passes easily into a craving for power or acclaim. So it is with the search for simplicity. The catharsis of insight is exhilarating, but the distinction between a genuine insight and a self-induced illusion is not clear. There is, however, one test to which one can put one's insights if one has the courage. Do they give one a feeling that they have satisfied the craving, or do they only make the craving more intense? Does the insight seem like an answer or even as "the" answer? If so, it is probably false. If it only opens the mind to further, more tantalizing questions, if it makes one more humble than proud, it may be genuine. Insights derived from speculations instigated by perceived analogies function somewhat like education: they reveal to the intelligent and conceal from the stupid the extent of their own ignorance.

Therefore, seek simplicity and distrust it.

2

The Evolution of Self-Simplifying Systems

HOWARD H. PATTEE

A few days before this Symposium, when I had the pleasure of meeting Professor Von Bertalanffy for the first time, he gave me a reprint of his paper, "Chance and Law,"[1] in which he presents several problems that are not adequately answered by the standard theories of biological evolution. I found this a stimulating paper, as well as reassuring, since the first problem Von Bertalanffy states is the one I had chosen to discuss at this Symposium in his honor.

The question can be stated in several ways: Why does life get more and more complicated? Why are complex organisms better adapted than simple organisms? Why do some species evolve rapidly and some hardly at all? One should be skeptical of the premise that complexity always results in better adaptation. It is very difficult to give an objective measure of either complexity or adaptedness. Almost no one will dispute that mammals are somehow more complicated than bacteria, but it is far less clear that mammals are better adapted. How do we go about determining adaptedness? If a species is well-adapted does this mean that selection pressures are low and we should not expect rapid evolutionary change? Or is evolutionary creativity an important component of adaptation?

I prefer to think of adaptation in terms of functional effectiveness, since this allows me to consider specific function which is easier to evaluate. Now we can ask the same question the other way around: Why is more effective function associated with more complex structure? Is this generally true? In most man-made devices it appears to be true. Our automobiles, airplanes, radios, telephones—even our pencils and toothbrushes—evolved from relatively simple beginnings to levels of complexity that only experts can understand. In most cases the function has been improved—at least according to some limited measure of effectiveness. But of course here we run into difficulty with artifacts because the designer, manufacturer, and consumer may have different measures. I reject the rechargeable battery-operated toothbrush in favor of the manual model, but on the other hand I prefer transistor radios to crystal sets. I suspect that in spite of certain functional advantages of the electric toothbrush, the older and simpler model will not become extinct.

Perhaps this analogy applies to natural functions, too. The environment has many measures of functional effectiveness, and certainly under many circumstances the simplicity and durability of the function carries more weight than specialization and precision of operation. This analogy can account for different *branches* of evolution for organs which perform the same basic function, but there is still the question of what specific conditions or challenges of the environment favor the branch of long-range increase in complexity, and increase in specialization. Does this branch of increasing complexity continue indefinitely? Or is there always some complexity optimum beyond which the organism is exquisitely functional but about to become extinct? One of the most discouraging properties of our technically proficient society is that while functions are maintained they may be quite elegant, but when failure occurs it is catastrophic. This argument has of course been used as one reason for extinction, although the opposite argument can also be used, i.e., that a needed function never evolved at all.

We find that the theory of evolution is more often criticized for its failure to make clear how new functions arise from random search rather than its complementary failure to make clear why almost all species become extinct. However, if in defending the theory, an evolutionary biologist invokes any optimality principle too generally, as often implied in "survival of the fittest," the theory is then criticized for explaining too much, i.e., being tautologous or untestable. One may therefore suggest, as Von Bertalanffy has done for many years, that some additional ideas or new approaches are needed if the theory of evolution is itself to evolve.

I have thought of the problem of evolution primarily in terms of the origin of life or of very primitive organizations. At first I accepted the problem as one of the growth of complexity from simple, and therefore probable, initial conditions. This is the form in which the problem is always stated, and it is the motivation behind many experiments and models.

The abiogenic synthesis experiments that have been performed show how the growth of complexity appears inevitable. One only needs the simple molecules of carbon dioxide, ammonia, methane, and water, and the energy of the sun, and spontaneous reactions produce most of the essential organic monomers and some of the polymers which are characteristic of living matter. In addition, of course, there is produced much insoluble, tarry material which is already too complex for chemical analysis.

I also tried to invent mechanisms for generating complex structures from simple rules, such as molecular representations of feedback shift registers for generating copolymer sequences;[2] but I found that the physical description of any device which executes a "simple rule" is exceedingly complicated. Von Neumann's lectures[3] on self-reproducing automata and Wigner's paper[4] on the impossibility of describing self-reproducing systems in quantum dynamical language stirred up my doubts that any picture of gradual and continuous growth of complexity in physical and chemical systems would eventually explain the origin of life. Von Neumann expressed his feeling that there must be some kind of Principle of Complication which says that unless some threshold level of complexity is reached, there can be no evolution of greater complexity; below this threshold systems degenerate. But perhaps more suggestive to me was his argument that self-reproducing systems which were capable of evolving complexity must have a self-description as well as self-constructing mechanism. This was really the first logical argument I had ever heard for the necessity of having both a genotype and phenotype for self-replication.

Wigner's paper, which I did not really understand or believe the first time I read it, pointed out that even though the Watson-Crick picture of template DNA replication and Von Neumann's picture of a self-reproducing automaton appeared to make good sense in the classical language they used to describe these pictures, they could not even be confronted with his quantum mechanical argument since none of the "tailored" constraints in these pictures could be expressed in the language of quantum mechanics. My lack of understanding of this paper led me to study the physical meaning of constraints and the physical difference between a description of an event and an event itself. This difference, it turned out, is at the root of the obscurity which still shrouds the interpretation of quantum theory, and is known as the measurement problem of quantum mechanics. It seemed to me very significant that the genotype-phenotype distinction which is essential for biological evolution, and hence for life as we recognize it, is also an essential distinction in quantum theory, and moreover, a distinction which is by no means clear.

There is a common requirement for both Von Neumann's self-reproducing automaton and any process which would be acceptable as a measurement, and that is a threshold level of complexity which *functions as a simple constraint* or rule. It then slowly dawned on me that the original program of searching for complex behavior from simple initial conditions had been turned around, whether I liked it or not, into the opposite program of trying to find simple behavior in initially complex systems. What do we do when we make a measurement? We want to record some single aspect of a system which potentially has many aspects. More generally we must classify a system into those few features which are relevant to our measurement and the many degrees of freedom that are not. Or to put it another way, what we measure is determined by what we choose to ignore about the dynamical details of a system.

In fact, the more I thought about it, the more it seemed strange that anyone trained in physics should regard the origin of complexity as anything but an unwanted, and all too common, property of nature. One could even characterize physics itself as the search for general simplicity in nature. And of course, much of this simplicity is of physicists' own making. They choose simple problems; but outside these problems, that is, beyond such abstractions as the single, isolated hydrogen atom, or a pair of point masses, the complexities reappear with a vengeance. The physicist must then "resimplify" the problem by choosing a new language which specifically ignores selected details. The scientific description of events would, in fact, get nowhere in nature's maze of complexity unless there were repeated resimplifications which we call hierarchical levels of descriptions.[5]

It also struck me as significant that the elementary event in the theory of automata or in switching networks is a selective loss of information or loss of some of the past history of the system. In fact every process of classification implies the grouping under a single class of objects which, according to a more detailed classification, fell into more than one class, i.e., the objects were initially further distinguishable. All symbols and records, insofar as they are classifications of events, are consequently simplifications of these events since they selectively ignore some details of these events.

Therefore it is quite reasonable to regard the origin of genetic records as a simplification of the underlying physical and chemical events which they represent. And if the basic necessary condition for the origin of life was the separation of the genotype from the phenotype—i.e., the separation of the description of the events from the events themselves—then it is fair to say that life originated by self-simplification of natural events.

How does this concept fit in with the conventional theory of evolution by random mutation and natural selection? Natural selection is the only known process in natural science for creating information.[6] Certainly the accumulation of this information in the course of time is the reason why species grow in complexity. Then where does simplification enter this theory other than at the origin of life itself? Once natural selection has taken over the course of evolution, is there any need for further simplification?

This question brings me back to the two strongest criticisms of mutation and natural selection as the only mechanisms in evolution. They are very old criticisms, but they can be found in a modern context in *Mathematical Challenges to the Neo-Darwinian Interpretation of Evolution.*[7]

The first criticism is that random search cannot be expected to "find" successful organizations because the search space is so immense. Thus a search for just one sequence of 100 amino acids representing one functional enzyme would require of the order of 20^{100} trials which is not within any reasonable probability within the age or size of the universe. Of course the numbers in this kind of argument are guesses, since we have no knowledge of the number of sequences which could perform a given catalytic function. However, the search space can also be enlarged almost without limit since we have no knowledge either of how many coordinated enzymes and other structures are necessary for self-replication. But at least no one should claim that there is enough experimental evidence either way to answer the question with any assurance.

The second criticism is that functional optima or fitness peaks in the adaptive landscape appear to be local, separate optima, so that no evolutionary pathway can be imagined that does not pass through nonfunctional or lethal valleys. Random search and selection can provide a mechanism for adaptation with respect to one of these optima, but for any fitness landscape with a realistic number of dimensions, the trapping problem becomes unsolvable. Reinforcing these general criticisms are the experiences of mathematicians with computers when they try to program them to show some reasonable degree of adaptability. In all cases, the immensity of combinatorial search space and the local trapping problems have not yielded to mutation and selection strategies.

The essential point is that both of these problems grow exponentially with the number of dimensions or degrees of freedom of the evolving system—that is, with the growth of complexity. Therefore, if mutation and selection are the only mechanism of evolution there are impassable limits to the complexity of organisms. Perhaps this is the case, but since we have already invoked a kind of self-simplification to explain the origin of life in the first place, and since we can actually observe this creative process at work in our many higher levels of description of nature, why should we not consider self-simplification as an evolutionary process at other levels in between?

I would like to propose that self-simplification is the key process in all so-called novelties or archetypes in evolution—also called transpecific or macroscopic evolution—as contrasted to local optimization for which mutation and natural selection appear to be a reasonable mechanism. Self-simplification does not compete with natural selection, but is complementary to it. That is, natural selection works well on relatively simple systems with the result that they accumulate genetic information and hence phenotypic complexity. This complexity is multiplied by the population's growth and stronger coupling with the ecosystem.[8] Consequently the organism reaches a level of complexity where natural selection stagnates and can only produce nonselective genetic drift. At this stage the more global process of self-simplification can introduce a new hierarchical level of description which selectively ignores the trapped details at the lower level. Natural selection can now begin to operate effectively in the language of this new simplified level; and the process can now repeat with no obvious limit to the level's structure.[9]

Can we say any more about how this self-simplification takes place? Can we test this theory in any way? Can we make a model that demonstrates this type of evolution? Can we expect a predictive theory of self-simplification? To begin with the last question, I would say that my picture of the self-simplification process would preclude any useful prediction of the detailed outcome. There are two arguments against detailed prediction. The first is that I see the simplification as partly a response to a physically intractable level of complexity, that is, a level which is quite beyond any artificial symbolic representation or calculation. For example, I do not believe that one can predict the catalytic specificity and power of a single polypeptide chain of 100 amino acid residues, given only the sequence, the solvent, and the equations of motion. The second argument is that the essential structure of the simplification, i.e., the hierarchical rules at the new level, are in the nature of frozen accidents, that is, they are largely arbitrary constraints operating on largely arbitrary symbol vehicles. This is the nature of all higher languages, and I would not be surprised if it were true of the genetic code. What we can hope to understand, however, are the essential properties of languages in the general sense of their function in the role of establishing new hierarchical levels of description.

This brings me to the question of testing this theory or making a model which demonstrates a self-simplification process. Here I believe there is already some suggestive evidence and good models, although this area of study is very new.[10] Perhaps the simplest demonstration of self-simplifying behavior are the randomly associated random switches of Kauffman[11] which have an enormous potential complexity in terms of the distinguishable number of states (for N switches there are of the order of 2^N states) but which spontaneously cycle over an average of only $\sqrt{N}$ states. This model could conceivably represent what would happen in a primitive reservoir of randomly specific catalysts where out of an immense number of catalytic sequences only short, stable cycles predominate.

A second, more mathematical view of self-simplifying processes is the topological dynamics of Thom.[12] Here the simplification results from the assumption that it is the discontinuities in the vector field of trajectories which determine the simplified structure at a higher level. Furthermore, these discontinuities may exhibit a kind of stability which is entirely deterministic in the evolutionary sense, but not predictable in the dynamical sense. It is also possible to make computer simulations of evolution which generate their own internal dominant behavior that is deterministic but not predictable.[13] Finally, I have suggested studying self-simplification in a real physical-chemical experiment such as the simulation of a primitive, sterile seashore where we watch for dominant behavior to condense out of the initial chaotic matrix.[14]

If we are to find out whether or not the evolution of higher organisms requires self-simplification processes in addition to mutation and selection, we must develop this type of study until we have a much clearer view. In any case, we see no way in which normal mutation and natural selection can account for the origin of life, or the origin of universal properties of life, such as the choice of amino acids, and the genetic code, and therefore we must take a broader look at evolutionary theory in the search for new principles.

3

Some Systems Theoretical Problems in Biology

ROBERT ROSEN

1. Introduction

Biology begins with the recognition of what we call living *organisms* as a separate class of entities, distinguished in structure and properties from the rest of the natural world. The intuitions on which this recognition is based are a mixture of introspections and experience, which despite great effort have never been completely formalized; that is, no one has ever been able to put forward a finite set of structural propositions that are satisfied by exactly those physical systems which our intuition tells us are organisms. Nevertheless, most of us take our intuitions on this matter seriously enough to believe that we can make a useful, scientifically significant distinction between living and nonliving, organic and inorganic. The absence of formalization means, however, that we cannot sharply specify the boundaries which separate the living from the nonliving. We encounter such boundaries when we ask, as some people do, whether viruses are alive, or whether it is possible to construct machines which can "live" in some sense, or whether there are other kinds of physicochemical systems (e.g., on the planet Jupiter) which we would want to classify as "living systems."

The absence of a formal characterization of a definite class of "living organisms" by means of a finite set of either-or propositions has long bothered biologists and philosophers of science. It seems to me that the difficulty arises simply from the fact that our biological intuitions are in fact not based primarily on the kinds of structural or metric considerations which, for example, dominate physics but are rather of a *relational* or *functional* character. The same difficulty of definition in fact arises whenever we try to specify in purely structural terms a class whose elements are defined functionally. To give one example; Wittgenstein asks:

45

How should we explain to someone what a game is? I imagine that we should describe *games* to him, and we might add, "This *and similar things* are called *games*."

Replace the word "game" by the word "organism" and we have exactly the biological situation.

We can see already that such relationally or functionally defined classes, and the intuitions to which they correspond, lean heavily on behavioral or dynamical analogies exhibited by the members of the class, or better, in the way in which we ourselves interact with the members of the class. And as we shall see, a study of such classes depends heavily on metaphors and metaphorical arguments. In the following remarks I shall sketch one manner in which we can attempt to arrive at some understanding of the behavior of systems which our intuition tells us are "living organisms"; how such understandings are to be related to our understanding of other areas of knowledge; and how the methods used to achieve this understanding can be applied to the study of other kinds of systems. This paper, then, constitutes an attempt to indicate at least some of the general system-theoretic concepts which are involved in attempting to construct a language suitable for the study of biological phenomena and biological organization.

Experimental biologists usually consider remarks of this kind to be of a very general character, and so in a certain sense they are. But in another sense they represent only a very limited and circumscribed area of application for these system-theoretic ideas. I have attempted in a few places to indicate some of the ways in which these ideas can be extended to other classes of organized systems, although the details of such extensions must be carried out by those with more detailed competence in these fields than I possess.

Any attempt to develop the general system-theoretic framework of even the limited area represented by biological systems must necessarily be permeated by the prophetic influence of Ludwig von Bertalanffy, and the present work is no exception. Indeed, the discerning reader will find on almost every page a clear indebtedness to Von Bertalanffy's pioneering work in general systems theory. The arguments I present against naive reductionism, the power of systems analogies in unifying apparently disparate branches of biology (and of systems in general), and the remarkable regulatory properties built into open systems, in particular, represent paraphrases of arguments Von Bertalanffy has expounded many times in the past, recast into the technical language which I believe most appropriate for drawing further specific biological inferences. The main claim to novelty in my own exposition, in fact, is in its promotion of the *intensive* technical development of system-theoretic tools appropriate to treat deep questions of biological structure and organization. This development should be viewed as a complement to the *extensive* approach most recently argued by Von Bertalanffy, which exhibits biological phenomena as merely one class of realizations of system-theoretic organizations, pervading the physical, engineering, and social sciences. Indeed, it is my belief that, now that the general conceptual framework of system theory has been laid (owing largely to the pioneering and heroic efforts of Von Bertalanffy), the next urgent task is to proceed to the parallel intensive study of many different classes of systems, aided and guided by the general homologies which the general theory of systems teaches us must hold between them.*

2. Generalities on the Modeling of Biological Systems

The only consensus found among biologists about their subject is that biological systems are complicated, by any criterion of complexity that one may care to specify. Therefore, if we are to achieve any kind of insight into the behavior of organisms, we must find some way of circumventing their inherent complexity; we must simplify them or abstract from them in

* The reader interested in exploring more deeply the relationships between the specific arguments developed below and the more general system-theoretic concepts which they illustrate, is referred to Von Bertalanffy's *General System Theory,* and for a more specialized treatment of biological systems, to his *Biophysik des Fliessgleichgewichts.*

some way; we must make models. Here the term "model" is to be taken in the widest sense; a molecular biologist, preparing a precisely definable fraction of the contents of the cell, is performing an abstraction; his resultant fraction is a simplified or abstract cell; it is a model. However, sharp controversies have arisen regarding the nature of biological modeling, the kinds of modeling which are acceptable, and the meaning and interpretation of biological models in this sense. At one pole we find the extreme reductionists, whose position will be considered in detail below; at the other we find holists who claim that any attempt to reduce the inherent complexity of organisms thereby automatically destroys their organic character, and that therefore any information pertaining to a model cannot pertain to the organism itself and must be erroneous.

A proper understanding of modeling in biology must, I feel, begin with an understanding of the interrelationships of physics and biology, which are profound and many-faceted. On the one hand, biological organisms are composed of atoms and molecules, and hence they simply *are* physical systems. The physicist is concerned with understanding the behavior of all assemblages of physical particles, including those that comprise organisms. And it is the fundamental principle of reductionism in biology that we have no real understanding of biological activities unless and until this understanding is expressed directly in terms of the interactions between the physical particles of which the organism is composed, i.e., in terms acceptable and recognizable to the physicist. This view, then, implicitly denies that there is any useful distinction between the organic and inorganic; between biology and physics.

A second and rather more subtle relation between physics and biology, which impinges even on holistic and systemic attempts to model biological systems, is that *the very machinery of system description,* the only tool we possess for this purpose, was developed for the analysis of simple physical systems (originating in Newtonian mechanics) and that despite extensive generalizations and refinements we still have no other conceptual tools available to describe systems and their behavior than those which proved convenient for physics.

A third relation, which plays a decisive though implicit role in motivating the reductionist viewpoint, is a counterpart of the preceding; namely that the only *experimental* tools available for the study of biological systems are also of a physical character. We have already mentioned that it is the manner in which we interact with systems which defines their character for us; in experimental biology we are constrained to interact with biological systems by means of techniques and tools invented by the physicist for studying inorganic nature. This bias on the manner in which we can observe biological systems automatically constrains us to a highly physical view of these systems, selectively emphasizing those aspects of biological systems which our observing procedures, drawn from physics, are geared to detect.

Thus both the experimental tools with which we observe biological systems, and the conceptual constructs by means of which we attempt to describe them, are drawn from a non-biological science, not concerned specifically with the complexity and the highly interactive character typical of biological organisms. Therefore, in order to orient ourselves properly with regard to understanding how the modeling of biological systems is to be effectively accomplished, we must understand more specifically the nature of the biases which our physical tools, both experimental and theoretical, impose on us. We therefore turn now to a discussion of these matters.

3. Systems and Their Descriptions

In both physics and biology, and indeed in all other sciences of systems, there are essentially two ways in which we can attempt to obtain meaningful information regarding system behavior and system activities. We can either passively watch the system in its autonomous condition and catalogue appropriate aspects of system activity, or else we can actively interfere with the system by perturbing it from its autonomous activity in various ways, and observe the response of the system to this interference.

In systems for which the passive, autonomous aspect is paramount, a kind of system description is appropriate which we shall call an *internal description.* Typically such a description begins with a characterization of what the system is like at an instant of time; such a characterization is said to define a *state* of the system. The totality of all the possible states of the system, meaning the totality of different aspects the system can assume for us at an instant of time, forms a set called the *state space* of the system. In physics these states are typically defined through the measurement of certain numerical-valued observables of the system; these are called *state variables,* and typically have the property that if two states are at all different in any observable way, they differ in the values assigned to them by one or more of the state variables; if two states are identical in the values assumed on them by the state variables, they are identical in all other observables as well.

In Newtonian mechanics it is a consequence of Newton's laws that a system consisting of N particles may be described by a set of only 6N state variables; out of the infinity of system observables these are conventionally taken to consist of three variables of spatial displacement for each particle in each of the three spatial dimensions, and three variables of velocity or momentum for each particle in the direction of the corresponding displacements. Thus the state space (or phase space) for such a system can be identified with a subset of ordinary Euclidean 6N-dimensional space, and each state with a point of this space. But it must be carefully noted that there is nothing unique about a set of state variables.

The fundamental problem of system description is to determine how the internal states change with time under the influence of the *forces* acting on the system. In physics such dynamical problems are formulated in terms of differential equations, which specify the rate at which each of the state variables is changing with time. The solution of a dynamical problem thus involves the integration of a set of differential equations, with each solution specified uniquely when the initial state of the system and the particular set of forces acting upon it are known. The temporal evolution of the system thus takes the form of a curve, or *trajectory,* in the state space.

The other kind of system description is called an *external* description, sometimes graphically called a *black-box* description. In this situation we make no attempt to identify a set of state variables for the system. Rather we have at our disposal a family of perturbations which we can apply to the system, variously called system *forcings* or *inputs,* and one or more observables which we use to index the effect of applying a particular forcing or input to the system. Such system observables are generally called system *outputs,* or *responses.* In general, in this approach, it is desired to determine what the system response will be to an arbitrary input.

These two approaches are, of course, closely related conceptually. By the way in which internal state variables are defined, any system observable (and in particular the system outputs) must be already a function of the state variables themselves. Each forcing or input in our repertoire must correspond to a set of equations of motion of the system, and hence the system response to each particular forcing can be calculated by integrating the corresponding equations of motion. But in dealing with any particular problem, it is often most cumbersome to try to find an appropriate set of state variables, and we can proceed simply by an input-output analysis without talking about state variables at all. On the other hand, given a particular input-output analysis, it is theoretically possible to formally find a set of state variables for the box itself; what these formal state variables mean is usually not obvious.

It is one of the goals of science to be able to match up the two kinds of system description we have described. The external description is a functional one; it tells us what the system does, but not in general how it does it. The internal description, on the other hand, is a structural one; it tells us how the system does what it does, but in itself contains no functional content. We would like to be able to pass effectively back and forth between the two kinds of system description; i.e., we would like to be able to infer the system function (the external description) from a knowledge of system structure (the internal description), and conversely, knowing the system function, we would like to be able to determine at least something about its structure.

In actual practice, theoretical physics is dominated by internal descriptions; the natural systems with which the physicist deals are generally of a simple type to which the concept of "function" is not appropriate. External descriptions begin to become important when we discuss artificial systems, especially the regulation and control of machines which we build for ourselves. Since in engineering we do things in the fashion simplest for us, our regulatory and control systems are related in a rather transparent way to corresponding internal descriptions.

In biology the situation is quite different, for a variety of reasons (some of which will be explored shortly). The crux of the matter is that a biological system is built on quite different (and largely unknown) principles from those systems which we build for ourselves, and our descriptions of organisms possess a curious mixture of internal and external characteristics. Many biological activities are in fact defined and observed only functionally, in terms of an input-output formalism. On the other hand, we can, as noted previously, employ many observational techniques (borrowed from physics) to obtain a wide variety of structural information. But there is no reason to expect that the structural information we find easy to measure should be related in a simple way to the external functional descriptions in terms of which so many biological phenomena are defined. Stated another way, the internal state variables which we find easily accessible bear no simple relation to the functional activities carried out by a biological system; and conversely, the external descriptions appropriate to the functional behavior of biological systems bear no simple relation to the structural observables which our physical techniques can measure. In the next few sections we shall explore some of the ramifications of this peculiar situation.

4. The Structural Characterization of Functional Properties

The remark closing the preceding section has an important bearing on the reductionist hypothesis, which asserts that the basic problems of biological systems can all be effectively understood in terms of the internal descriptions of physics (using as state variables the observables defined through the use of observing systems likewise drawn from physics). The question is then: how does a physicist approach a physico-chemical system too complex to be studied as a whole? As indicated previously, he must abstract from or simplify the system in some way. The customary way is to physically fractionate the system; break it up by physical means into a spectrum of simpler subsystems, if necessary iterating the process by fractionating the individual fractions, until we are left with a family of subsystems each of which is simple enough to be studied as a whole. He then will attempt to assemble the information he has obtained regarding each of the fractions into information about the original system with which we began. Implicit in this are two crucial hypotheses, of a system-theoretic character: namely, that any physicochemical system, however complex, can be resolved into a spectrum of fractions such that (a) each of the fractions, in isolation, is capable of being completely understood, and, most important, that (b) *any* property of the original system can be reconstructed from the relevant properties of the fractional subsystems.

This last hypothesis is demonstrably false for many systems, including most of those of biological interest. A simple physical counterexample is a system of three gravitating masses in space (three-body problem). We can surely fractionate a three-body system into various two-body and one-body systems, each of which is simple enough to be completely understandable in isolation. But the crucial stability properties of a general three-body system can never be reconstructed from a knowledge of two-body or one-body systems, however comprehensive. The basic reason for this is that the fractionation techniques employed are not compatible with (or do not *commute* with) the dynamical properties of the original system; we irreversibly destroy this dynamics, the very object of interest, by the process of fractionation itself.

Thus when we apply a prespecified set of fractionation techniques to an unknown system, there is no reason why the fractions so obtained should be simply related to properties of the original system. Yet this is exactly what happens when a molecular biologist fractionates a cell and attempts to reconstruct its functional properties from the properties of his fractions.

This is not at all to say that fractionation per se will give no information about the properties of a complex system (although holists will go that far). What we must do to accomplish this is seek fractionations compatible with the system dynamics, in a definite, well-defined sense. We have argued that the fractionation techniques imported into biology from physics will not in general be compatible with the dynamics of biological systems. This does not imply that such fractionations do not exist; they may well exist, but they will generally be of a different character from those which have heretofore been important in analytical biology. They will be, in some sense, "function-preserving," as in the following simple example. We all know that a bird's wing is a combination propeller and airfoil, with both functions inextricably intertwined. This is different from the case of an artificial system like an airplane; we can physically fractionate an airplane into physically distinct parts which preserve such functions, but such procedures fail in the case of the bird's wing.

This example illustrates on one hand the different principles of construction on which biological structures and engineering structures are built, and at the same time illustrates that the fractionation procedures appropriate to biological organization must be of a different character than those appropriate to simpler physicochemical systems. Basically this is because in biological systems the same physical structure typically is simultaneously involved in a wide variety of functional activities.

The situation with regard to physicochemical fractionations of arbitrary systems is actually much worse than this, as will appear in the next section. But we have already shown enough to demonstrate that a simple reductionist hypothesis cannot be true for at least many of the functionally defined properties of the greatest biological interest.

5.   System Analogies

As I said in the preceding section, a set of structurally meaningful state variables for a biological system is most difficult to identify, particularly if we restrict ourselves, as we usually do, to those quantities defined by purely physical observation techniques. We may always have recourse to an external description, i.e., to an input-output analysis; this is always appropriate to a system defined primarily in functional terms to begin with. But such black-box descriptions, though they are very useful (and allow us to make predictions about our system) carry with them only a limited understanding. Only an internal description, or something very much like it, can allow us to say that we fully understand the behavior of our system.

There is a sort of halfway house between internal and external descriptions which allows us to go a bit further than we can with external descriptions alone; this depends on the concept of system, and what I have called elsewhere the construction of dynamical metaphors for biological activity. Let us begin with the notion of analogue, which has long been employed by experimental biologists in the study of complex systems, under the generic term, *model systems*. Thus we find

55

enzymologists attempting to learn about enzymes by studying systems ("enzyme models") which are not enzymes; we find physiologists attempting to learn about the properties of biological membranes by studying collodion films, thin glass, artificial lipid bilayers and other types of "model membranes"; we find neurophysiologists attempting to learn about the nervous system by studying a variety of artificial switching mechanisms or other forms of "neuromimes"; and all kinds of scientists attempting to understand the dynamics of their system of interest, whatever its character, by modeling on an analogue computer, i.e., an electrical system so constructed as to mimic the original dynamics.

The reason that the use of model systems is possible at all is that the same dynamical or functional properties can be exhibited by large classes of systems, of the utmost physical or chemical diversity. Two systems which are physically different but dynamically equivalent will be called *analogues* of one another (the terminology obviously drawn from analogue computation, which embodies this concept in a particularly transparent way). If our interest is in the system dynamics, then this dynamics can be studied equally well (and often better) in any convenient system analogous to our original system.

Modeling by system analogy has obviously a completely different basis than the kind of fractionation we discussed in the preceding section. System analogy shows us that dynamical or functional properties can be studied essentially independently of specifics of physicochemical structure, while fractionation, or other reductionist techniques, are bound up with these specifics in an essential way.

Analogies of this kind are common even in theoretical physics. The mechano-optical analogy of Hamilton and Jacobi or more generally the organization of whole branches of physics around analogous variational principles is well known. Indeed, the judicious exploitation of such variational principles is one of the most impressive unifying agencies which exists in physics, potentially binding all of physics together in terms of functional or dynamical analogies, instead of attempting a unification on the basis of the structural fact that every physical object is built out of the same set of elementary particles. We shall see that the concept of system analogy plays an equally striking unifying role in biology. Such a unification is the sole attractive aspect of biological reductionism; I shall suggest that one can hope to achieve unification on functional terms while avoiding reductionist pitfalls.

The concept of system analogy is a most interesting one mathematically, and even opens up new vistas in classical physics. System analogy is most conveniently defined in terms of internal descriptions; two systems are analogous if, roughly, there is a 1–1 mapping between their state spaces which commutes with the system dynamics. But in physical systems, we have not only the state variables, but the full set of system observables (i.e., real-valued functions on the state space) available to us. Once a set of state variables, and the equations of motion, of a system are specified, every observable of this system inherits a particular dynamics. It is generally possible to find sets of such observables which define dynamical systems in their own right—such systems are in effect subsystems of our original system. It turns out in fact that there exist physical systems which are *universal* in the sense that we can build a dynamical system out of appropriate observables of the universal system which is analogous to *any* arbitrary dynamical system.

This kind of result has many profound implications. For one thing, we have already mentioned that we apprehend a system in terms of those system observables which are in some sense easy for us to measure. The same system would present itself to us quite differently if we interacted with it differently; i.e., if other observables of the system were made easy for us to measure. Indeed, a universal system could be made to appear as any arbitrary dynamical system, simply by interacting with it in an appropriate way. This may open novel possibilities for simulation. And, returning to the notion of fractionation: we can fractionate such a universal system in such a way that the isolated fractions have *arbitrary* dynamical properties. This shows in a particularly graphic way the difficulties inherent in attempting to infer system dynamics from a study of fractional subsystems separated by conventional physicochemical means.

Let us return to the statement made previously that any functional or dynamical property of a given system can be studied equally well on any one of the system analogues, or even entirely in the abstract. Such an abstract functional property, exhibited by each of the system analogues which *realize* the abstract system, is what we call a "dynamical metaphor." For example, there are a number of important biological properties which follow simply from the fact that biological systems are open systems in the dynamical sense. To understand such properties we do not need to know which open system, in complete structural terms, is in fact before us, but merely that it is open. This is a situation familiar even in mathematics; if a particular property of a group, for example, follows simply from the group axioms, then it is redundant, and indeed incorrect, to prove the result by invoking the specific properties of the group elements comprising the specific group before us. In this way we can begin to carry out what we may call "functional fractionations," which allow us to see what follows already from the simplest dynamical properties of a metaphor, and what properties require the invoking of more specific dynamical or structural assumptions.

Such dynamical metaphors are playing an increasingly important role in our understanding of biological processes, different as they are from conventional structural modeling. The use of model systems has already been mentioned, as has the employment of open systems as metaphors for switching systems, threshold elements, equifinality in development and regeneration, etc. Another popular dynamical metaphor is the employment of a single metastable steady state as a metaphor for the establishment of polarities or gradients in differentiating systems.

There is one difficulty in the study of dynamical metaphors which must be mentioned. A dynamical metaphor, by its very nature, refers to a *class* of analogous systems, which may be of the utmost physical diversity. A typical biologist, on the other hand, is interested in the specific system before him, and asks for specific structural implications of any theoretical scheme that he may test on his system. Obviously the dynamical metaphors are not, by themselves, geared to provide us with specific structural information about individual systems in the class. Thus it is difficult to make explicit contact with the structural information available about individual biological systems, which after all comprises the vast bulk of our biological knowledge. We have suggested elsewhere that dynamical metaphors, appropriately supplemented with further conditions (in particular, with constraints arising from considerations of optimal design) allow us to pick individual systems out of a large class of analogous systems (namely those which satisfy the additional constraint of optimality), and about these individual systems we can make a great many more specific structural inferences.

6. Hierarchical Systems in Biology

I have stated several times previously that biological systems are constructed along different principles from the simple physical systems and engineering artifacts with which we are most familiar. One of the most obvious of these differences is the pronounced hierarchical character of biological systems; the separation of biological activities into distinct levels of organization. I shall call a system *hierarchically organized* if it satisfies the following two conditions: (a) the system is engaged simultaneously in a variety of separate distinguishable activities, and (b) different system descriptions are necessary to describe these several activities. It is this second condition which characterizes biological systems, with their stratification into many levels of organization.

We must say a word about what is meant by "different system descriptions." We pointed out above that the same system always admits at least an external description and an internal description, and that these are different. However, this is not the kind of difference which is meaningful for hierarchical organization. What is meant is that the system requires several essentially different internal descriptions (each of which carries with it a corresponding external description) to account for the various activities of the system. A simple physical example should make this clear. We can regard a gas in two quite different ways: on the one hand it can be regarded as a structureless fluid, describable in terms of the thermodynamic state variables (pressure, volume, temperature, etc.). On the other hand, a gas can be regarded as a very large number of small Newtonian particles, admitting an internal description in terms of the state variables appropriate to the dynamics of Newtonian systems; displacements and their corresponding velocities or momenta. These two state descriptions are essentially different; they refer to different structural levels of organization of the gas, and are made apprehensible to us in quite different ways.

Biological systems are very highly stratified in this sense, into levels ranging from the submolecular to the ecological. A great deal of theoretical biology in the past was devoted to an attempt to find an "anchor" level in the hierarchy; a level which was biologically meaningful, understandable in its own terms, and most important, would allow us to infer the properties of all the levels above it and below it in the hierarchy. For many years it was thought that the cellular level was such a level; this was the deeper significance of the cell theory of Schleiden and Schwann. With the advent of biochemistry and molecular biology many biologists regarded the biochemical level as the most appropriate "anchor" in the hierarchy. Indeed, the essential content of the reductionist hypothesis was that it asserted that it was possible to infer the properties of any level in the hierarchy from the biochemical level.

In the preceding sections we have seen some of the acute problems of the reductionist hypothesis; here we consider the

60

question of whether it is, in fact, possible to pass *effectively* from the biochemical level to higher levels of biological organization. This is a question which has received ample attention in recent years, from such authors as Elsasser, Wigner, Polanyi, Pattee, and others. This is not the place to go into specifics of these arguments; it need only be stated that it is at best exceedingly problematic whether one can indeed effectively traverse organizational levels when one starts at the bottom. Even in physics, the tool used for passing between the dynamic and thermodynamic descriptions of a gas is statistical mechanics, a tool of the greatest difficulty and subtlety, which has hardly been fully mastered and is at best of limited applicability. And although it seems on the surface that statistical mechanical ideas can be readily imported into biology, the several attempts to do so have run into the gravest technical and conceptual difficulties.

There is, however, one aspect of the hierarchical organization of biological systems which bears mentioning at this point. Namely, it appears that the dynamical properties which emerge at successively higher levels of biological organization are *analogous,* in the strict sense employed in the preceding section, to those at the lower levels. One particularly striking instance of such analogies occurs between the biochemical and genetic control networks found by Jacob and Monod, and the neural networks in the central nervous system. The exploitation of this analogy may ultimately prove as fruitful for biology as the mechano-optical analogy has been for physics.

7. Implications for "Structural" Studies of Complex Systems

The main points which have been made in the above discussion are:

a. That the only way in which we know how to approach complex systems is to simplify or abstract from them in some way;

b. That such simplification amounts to splitting our system

61

into subsystems, which are simple enough to be characterized in isolation, and such that our knowledge of the isolated subsystems can be effectively employed to give us information about the original system;

c. That in biology, the abstractions offered by physical reductionism do not in general satisfy proposition b, in that they are not generally compatible with the dynamics of the original system.

We believe that the first two of these propositions are universally applicable to a study of complex systems of whatever type; social, economic, political, linguistic, etc. What we seek in the study of such systems is a spectrum of "atomic" subsystems which can be understood in isolation, and whose essential properties are preserved when a set of such "atomic" subsystems are recombined. What point (c) above tells us is that we must avoid preconceptions as to the nature of such "atomic" subsystems; that those subsystems which seem a priori to be the most natural candidates for this purpose may in fact not be so; and that we must let the overall system dynamics decide this for us.

Actually, the identification of such "atomic" subsystems implies far more than this. For, by their very nature, such subsystems can be juxtaposed or recombined to produce new kinds of systems, different from those with which we originally started. That is, we may use these subsystems as elements from which new kinds of systemic organization can be synthesized by a set of canonical rules for the juxtaposition of our "atomic" subsystems. In such a situation, the systems with which we started are displayed as special cases of a generally much larger class of systems, all constructible from a family of "atomic" subsystems by means of a definite set of formal rules of combination or juxtaposition. This kind of analysis followed by resynthesis is typical of many fields within pure mathematics and the applied sciences; it is for example the basis of the numerous "canonical form" theorems for algebraic or topological structures.

It often happens, however, that we wish ultimately to restrict our attention to those synthetic reconstructions of our "atomic" subsystems which correspond to "natural" objects. We thus desire a set of rules which can characterize, out of the class of all such synthetically reconstructed systems, the subclass of those which are "natural." This is in general a much harder problem even than isolating our "atomic" subsystems in the first place. It amounts to exhibiting a set of rules (a "grammar," if you like) whereby the natural systems can be effectively exhibited, or at least effectively recognized. We pointed out the difficulties in carrying out such a program for biology at the outset of these remarks, when we noted that the class of "natural" biological organisms has never been successfully characterized within the class of all physical systems (implicitly taking for our "atomic systems" the set of real physical atoms). But this does not mean that we can never carry out this specification with *any* choice of atomic subsystems. We know already, indeed, that for biology the choice of such subsystems will be quite different from those the reductionist hypothesis gives us.

The entire process we have just sketched, beginning with the isolation of atomic subsystems, their recombination to generate a large class of systems, and the rules for selecting a subclass of systems of interest out of these, are implicit in the notion of "structuralism" or structural analysis for the study of complex systems. We have amply seen that the word "structural" has to be interpreted in a very wide sense; indeed in biology the relevant "structures" are always defined in *functional* terms.

It may be helpful to itemize the procedures involved in such a "structural analysis" of biological systems. This itemization is rather complicated, but once the essential aspects are systematically set down it will be recognized that exactly the same procedure is implicit in the structural study of all other kinds of organized complex systems.

We begin by supposing that we have already identified a

class of "atomic subsystems" satisfying the hypothesis point (b). We may as well suppose that these are completely abstract systems, because by hypothesis any real biological system can be decomposed into real subsystems which realize such atomic subsystems; but in general different biological systems will give us different (but analogous) atomic subsystems realizing the same abstract systems. Let us designate this set of abstract atomic subsystems by the symbol A.

We now suppose that these abstract atomic subsystems can be combined or juxtaposed by a definite set of canonical operations or rules of composition, to form a large set of abstract systems, which we may suggestively designate as *abstract words,* and denote as $A^\#$. $A^\#$ is thus the set of abstract systems *generated* from A by the employment of the canonical composition rules.

Finally, we wish to identify or select out of $A^\#$ a subset, B, corresponding to the "abstract biological systems." The words of $A^\#$ not in B are the "abstract nonbiological systems." The set of rules we use to make such a selection or identification of the elements of B we may suggestively call a "grammar."

We thus have a sequence of operations going from the set A of abstract atomic systems to the set $A^\#$, the set of abstract words, to the set B (the set of abstract biological systems), which may be represented by the following diagram:

$$A \xrightarrow[\text{rules}]{\text{juxtaposition}} A^\# \xrightarrow[\substack{\text{rules} \\ (\text{"grammar"})}]{\text{selection}} B.$$

Now the set A of abstract atomic systems can in principle be *realized* in physical terms in many different ways. Suppose that such sets of specific realizations are designated as

$$R_1(A), R_2(A), \ldots, R_n(A), \ldots.$$

For each i, the real systems in $R_i(A)$ realize the abstract systems in A; hence there is a natural mapping of A into $R_i(A)$

64

associating to each abstract atomic subsystem its realization, and a natural mapping of each $R_i(A)$ into each $R_j(A)$, associating to each system in $R_i(A)$ its analogue in $R_j(A)$.

The rules of juxtaposition of abstract atomic subsystems, by which $A^\#$ is generated from A, may now be realized in terms of specific physical operations or processes in each $R_i(A)$, *perhaps in many different ways*; i.e., using different physical processes to combine the systems in $R_i(A)$. Thus in general each $R_i(A)$ can give rise to many sets of juxtaposed systems or words, which we may designate as

$$R_{i1}{}^\#(A), R_{i2}{}^\#(A), \ldots, R_{ik}{}^\#(A), \ldots.$$

Each element of $R_{ik}{}^\#(A)$, for all i, k, is a realization of some word of $A^\#$; hence there is again a natural mapping of each of the sets $R_{ik}{}^\#(A)$ into each of the others which associates analogous words (two words being analogous if they realize the same word in $A^\#$).

Further, we may identify in each $R_{ik}{}^\#(A)$ a number of sets $B_{ik1}, B_{ik2}, \ldots, B_{ikj}, \ldots$, these being selected according to different physical "grammars" on the set of words $R_{ik}{}^\#(A)$.

We thus have many different candidates for "real" biological systems, specified by the diagram

$$R_i(A) \longrightarrow R_{ik}{}^\#(A) \longrightarrow B_{ik1}.$$

All such diagrams are connected by mappings into every other such diagram, which identify analogous but physically different systems. Presumably the study of "real" organisms is just one of these; whether the other diagrams are equally real (i.e., whether we can realize biological organization with novel physico-chemical structures), or whether other such diagrams are excluded on some kind of physical grounds, is an open question.

This formalism applies equally well to other kinds of organization, even nondynamical ones like linguistics. Here we can assume that there is only one set $R_0(A)$ of realizations of the set A, comprising the linguistic "atoms" (morphemes or phonemes), and only one rule of juxtaposition leading to the set $R_{oo}^{\#}(A)$ of linear sequences of linguistic atoms. But there are in general many "grammars" leading to different but analogous sets $B_{ooj}(i = 1, 2, \ldots)$ of "natural languages."

8. Evolutionary Problems

Before concluding this brief paper, it is necessary to add a further word regarding the evolution of biological structures in time. I have, in the preceding analysis, been concerned entirely with "physiological processes," those which take place during the lifetimes of single organisms. We have neglected developmental problems, and most particularly, we have neglected evolutionary problems, which are concerned with the way in which the class of organisms changes over long periods. Since a "structural" analysis pertains only to the class of biological systems at single instants of time (i.e., is a static description of the biological world, considered in evolutionary terms) there is an essential dynamical element missing from our discussion; in the terms used above, we have specified the instantaneous states of the biological world, but not the forces acting on them to produce changes of state, nor the equations of motion to which these forces give rise.

The way in which such equations of motion, corresponding to evolutionary processes, can be constructed and investigated is a vast and difficult problem, somewhat simplified in biology by the curious analogies which exist between evolutionary and developmental processes. In purely descriptive terms, evolutionary processes can be regarded as a temporal dependence of the "grammatical" rules whereby a set B is selected from the set $A^{\#}$. But such temporal dependence requires its own kind of "structural analysis," and how to go about making such an analysis in any kind of evolutionary situation is, to my knowledge, a completely open question.

4

The Impact of Von Bertalanffy on Physiology

EKKEHARD ZERBST

I N the following remarks I shall try to outline the conceptual constructs of Ludwig von Bertalanffy's leading ideas with reference to their impact on physiology. The importance of his work was acknowledged very early. In 1933, E. S. Russel stated: "Bertalanffy is one of a small band of people, who are paving the way to a new conception of the organism, a new orientation of biological thought."[1] And Joseph Needham noted: "Recognizing it as something new in biological literature, biologists everywhere will warmly welcome Dr. Bertalanffy's new work on theoretical biology. . . . Such a synthesis has never been attempted."[2] But the fruits of Von Bertalanffy's work, growing up from the seed sown in physiology and biochemistry, have been ripening in the last decade. His basic ideas in the organismic conception and in open systems theory have greatly influenced experimental methods and the quantitative handling of physiological problems especially in recent years. The organismic concept is now incorporated in the classical and somehow anonymous pool of biological knowledge. Organismic biology has not been replaced by modern molecular biology. As shown by Dobzhansky,[3] Commoner,[4] and Dubos,[5] the organismic concept proves the importance of molecular biology. And in the sense of Kuhn's *The Structure of Scientific Revolutions,* Bertalanffy's organismic concept, which has been incorporated in the modern theories of ecosystems, is a new paradigm. It is a new principle of orientation, developed in sharp disputes and controversies.

Bertalanffy's laboratory work has been in cellular and comparative physiology of metabolism, animal growth, cyto- and histochemistry and related fields. His laboratories in Vienna and Ottawa were among the first to investigate the quantitative physiology of metabolism on a broad comparative scale, especially in different invertebrate classes, and in connection with cell metabolism and growth. This was followed by many investigations of colleagues and former students—Needham, Allen, Krüger, Scharf, Fabens, and many others.[6] Using the basic laws of his organismic concept, Von Bertalanffy since 1933 developed his equations of organic growth, taking into account the continuous process in which both the so-called building materials as well as energy-yielding substances are regenerated and broken down in the open system of organismic and cellular structures. Today, the "Bertalanffy growth equations" are an important method of predicting the yield of fisheries all over the oceans, as recently shown by Beverton and Holt.[7]

Bertalanffy's advocacy of fluorescence microscopy in the study of animal tissues has been another practical consequence of his studies on growth. His work has drawn wide attention, particularly with respect to the acridine-orange method in cancer screening and research, differential diagnosis of malignancies, embryology, and virus research. This method is routinely and internationally used in hospitals, cancer clinics, and research institutes; more than 300 published clinical and research reports followed the initial publications by Von Bertalanffy and co-workers, including those of his son, Felix D. Bertalanffy.[8]

As M. N. Meissel states, . . . "It is not surprising that so outstanding a figure in theoretical biology as L. von Bertalanffy should devote considerable time and attention to cytochemistry of nucleic acids. These studies proved of fundamental importance to cytology and cytochemistry and led to the elaboration of fluorescence-microscopy of practical importance for detecting cancer cells, now widely used in clinical practice."[9]

It was in 1940 that Von Bertalanffy published his general concept of the kinetic theory of thermodynamically open living organisms. Using this it was possible to handle fluxes, driving forces, and steady-state transitions of matter and energy through the organismic system. In previous times, classical thermodynamics exerted a profound influence on biological thought. It succeeded in inserting living processes into the frame of the first and second laws of thermodynamics and assigned energetic and entropic quantifiers to metabolic processes. But the limitation of classical thermodynamics for the interpretation of organismic behavior has been due to its nature as an equilibrium theory. These states of equilibrium are "memoryless" (Katchalsky); dead ends of dynamic processes. The basic principles of Von Bertalanffy's open-system theory of 1940 have been promisingly advanced during the last decade through the application of nonequilibrium thermodynamics to biological processes. Latest results have been shown by Katchalsky and Spangler, Prigogine, De Groot, and Glansdorf.[10]

Nonequilibrium thermodynamics, however, is restricted to linear processes, proceeding mainly in homogeneous systems; in these Onsager's symmetry is obeyed, and an integration of the dynamic equations becomes possible. Since most metabolic processes are nonlinear, and all cells and tissues are nonhomogenous, a new approach based on modern network theory has been developed by Oster, Perelson, and Katchalsky.[11] Thus the thermodynamics of biological networks now permits a quantitative treatment of complex systems, composed of nonlinear elements in which processes far from equilibrium take place.

Von Bertalanffy's theory of open-system kinetics has been applied by B. Chance in important studies on special problems of cell metabolism.[12] By this the transitions of enzymes and co-factors under changing conditions from aerobiosis to anaerobiosis and from rest to activation have been discovered. B. Chance succeeded in giving insights into the possibilities of control in cell metabolism. By means of the formulation of the cross-over theorem—which is an exploitation of Von Bertalanffy's open-system theory—the transient processes have been formulated in rules. In the same way—starting from the open-system theory—Hess succeeded in proving the rules of permanent nonlinear oscillations of allosteric control enzymes in glycolysis, thus giving insights into the possible analogous basic mechanisms of the so-called "biological clock."[13] This was done some time before the controlling nodes of metabolism were investigated by kinetic methods, using the difference between the chemical equilibrium constant and the nonequilibrium constant.

In 1953 Von Bertalanffy noted that the mechanisms underlying the response of biological sensory cells to stimuli should be analyzed on the basis of calculations of the sensory cell metabolism. Our group (Zerbst and Dittberner) succeeded in performing this analysis.[14] We have shown that the stimulus-response characteristics under various conditions follow the same basic rules in sensory cells as steady-state transitions of metabolism in simple open systems when velocity coefficients have been influenced by some sort of stimulatory process. Starting from considerations of the stimulus-response dynamics of the simplest open-system models, the authors have shown how, by means of the method of open-system thermodynamics, it is possible to understand integrated metabolic systems as functional units. Step by step the model of the sensory cell metabolism was approximatively extended to the dynamics of ionic fluxes and active transport across receptor membranes and the correlated membrane potentials. In this way we made some tentative progress toward the most interesting connection between energetic entropy and the negentropy of information. Thus the heuristic value of Von Bertalanffy's theory was demonstrated, providing an example of his statement: "From a more general viewpoint, we begin to understand that besides visible morphologic organization, there is another invisible organization resulting from the interplay of processes determined by rates of reaction and transport."

One of these processes is, of course, the basic mechanism of biological information uptake and information conversion into the code of action potentials. At the least we have shown that system models—and especially electronic analogues which have been used in these studies—provide other approaches besides physiological experiment. They permit (1) solutions of multivariable problems which otherwise exceed time limits and available mathematical techniques in the theoretical branches of sciences; and (2) the application of electrical analogues of sensory cells as "artificial organs" in medicine. By means of the latter we succeeded in showing, for instance, that electronic receptor analogues can be implanted instead of insufficient baro-receptors regulating high blood pressure in hypertensive patients.[15] Thus, starting with considerations of the heuristic value of Von Bertalanffy's open-system theory for the solution of problems of biological receptor performance as information-converting systems, we arrived at the point of being able to demonstrate the clinical relevance of open-system analogues as a therapeutic method.

Further examples for the application and development of theory of open systems have been given in recent works on the development of systems theory by Rosen, on compartment theory by Rescigno, on the application of irreversible thermo-dynamics to problems of active transport by Katchalsky, on pharmacodynamics by Dost, and on the methods of open-system cell-culture by Malek.[16]

Modern investigations of metabolism and growth, physiology and biochemistry, psychology and psychiatry, all have to take into account that the living organisms, as well as their components, are so-called "open systems." Thus, in general, there is no field in physiology and biochemistry in which investigations would not have been influenced by the organismic concept and by open-system theory.

Last but not least, it must be mentioned that Von Bertalanffy's general systems theory has highly influenced the fields of neuro- and sensory physiology in the present decade. This has been fully recognized recently, especially in Europe. The program and principles of this theory had already been developed when cybernetics and mathematical systems theory were established. The influence of Von Bertalanffy's work in this field of modern science may be illustrated by citing two recent textbooks: W. Buckley, *Modern Systems Research for the*

Behavioral Scientist (1968), and F. E. Emery, *Systems Thinking* (1969). In behavioral physiology, psychology, and psychiatry, he likewise influenced the methodology of research. He attacked the robot model of man with particular vigor. The introduction of systems theory in psychology and psychiatry by Von Bertalanffy has been characterized by R. R. Grinker (1967) as "the third modern revolution" in psychiatry, following those of psychoanalysis and behaviorism.

Ludwig von Bertalanffy has been one of the few pioneers in the development of biological science, giving the basic concepts and starting points for recent investigations. He introduced theoretical biology as a scientific discipline. He was the first to whom, in 1934, a professorship of theoretical biology was given at the University of Vienna. As Professor Bavink stated in his *Ergebnisse und Probleme der Naturwissenschaften* (1933): "If anywhere a chair is to be established for the science of theoretical biology as the first of its kind, then Ludwig von Bertalanffy is one of the first who has a claim to having it . . . and it is really time that such chairs be established."

The topological point of biology (and, by this token, of physiology and medicine) within the whole system of sciences has achieved its significance by the fact that there is always some sort of "anabolic flux" entering from the "source" of physics and chemistry, the "open system" of biological science. On the other hand, there is also a flux of substantial and heuristic concepts, ideas, and methods, leaving the open system of biology and entering as steady advances and catalytic influences in the behavioral sciences, psychology, psychiatry, and sociology.

Scarcely any other scientist has had such an impact on theoretical biology as Von Bertalanffy in influencing as well as converting the concepts of classical physics and chemistry. It is simply a logical consequence that Von Bertalanffy catalyzed the conceptual fluxes from biology to psychology, psychiatry, and the behavioral sciences. This consequence can only be understood in its specific significance by taking into account his personality.

Von Bertalanffy's scientific history appears to us, beyond most others, as having a rich, subtle, and manifold significance. His life exhibits traits such as seldom occur in the history of science, and indeed seldom can occur. Rising in early manhood, almost at a single bound, into the highest reputation in the scientific world, he has continued to establish his position more and more firmly in the esteem of scientists. Today, after several decades of convulsions (scientific, philosophic, and political), he still labors in his vocation, still transmits with the benignity of a great man whatever can profit science. Supreme achievements of this sort are rare in modern times.

Let me conclude with Robert A. Smith:

Von Bertalanffy's early and fortuitous association with other highly distinguished men, such as Rapoport, Boulding, Gerard, Miller, and Ashby, provided an association of equally strong egos with no single one of this group seeking, or depending on, charismatic associations. This significant association reminds us of the association of Jefferson, Madison, Franklin, and Adams, or of people who surround themselves with strong but democratic dissenters, so that their collaborative assent breathes the power of process rather than the whims and structure of an idiosyncratic specialist who needs flattery. Bertalanffy's ego is powerful, but this has enabled him to invest it securely in the world exchange bank of knowledge and to sincerely praise those who advance general systems theory to more fruitful applications for science and in this way for mankind.[17]

5

Economics and General Systems

KENNETH E. BOULDING

I N my own recollections the Society for General Systems Research, as it later came to be called, originated in a conversation around the lunch table at the Center for Advanced Study in the Behavioral Sciences in Palo Alto, California, in the fall of 1954. The four men sitting around the table who became the founding fathers of the Society were Ludwig von Bertalanffy, Anatol Rapoport, Ralph Gerard, and myself—a biologist, an applied mathematician and philosopher, a physiologist, and an economist. Economics, therefore, can certainly claim to have been in at the beginning of that enterprise, although this may have been largely an accident of my own personal interests. Certainly one cannot claim that the interaction between general systems and economics has been very extensive since that date, though the contributions of each to the other may be more than many people recognize. In the intervening years, however, the social sciences in general systems have been represented more by sociologists, such as Buckley, and psychologists, such as the late Kenneth Berrien. Almost the only other economist I can think of who has played much of a role in the development of general systems is Alfred Kuhn, whose interest, like my own, has been primarily in going beyond economics to developing an integrated social science.

Just why the economics profession has viewed general systems with such a massive indifference I really do not know. Like the physicists, the economists are so bound up within the elegant framework of their own system that they find it hard to break out into a broader interest. Economics, indeed, may be a good example of a principle I have sometimes enunciated, that "nothing fails like success." The very success of economics, and especially of econometrics, in formulating systematic quantitative theories and methodologies may have prevented the profession from looking outside its own boundaries for further insights and models. I regard this as unfortunate, as in my view certainly the general systems approach to knowledge has important contributions to make to economics, as it does to virtually all other fields.

In the initial manifesto of what was then called the Society for the Advancement of General Systems Theory, which was published in the program of the Berkeley meeting of the American Association for the Advancement of Science in December 1954, a "general system" was defined as any theoretical system which was of interest to more than one discipline. On this criterion many and perhaps all of the theoretical systems of economics would qualify as general systems, for they are certainly relevant to other disciplines. The theory of the general equilibrium of the prices and outputs of commodities, for instance, as originally developed by Walras, and made in part operational by Leontief in his input-output analysis in the 1930's, is clearly a special case of a general system of the utmost importance, for it is a special case of the general equations of ecological equilibrium. In the simplest formulation of this system, we suppose a number n of interacting populations, each composed of the individuals belonging to a single species. The species here may be biological species, such as hummingbirds, or commodity species, such as automobiles, or even mineral species, such as available nitrogen in the soil or mineral nutrients in a pond. They may also be psychological species, such as the demand functions of different individuals for different commodities.

The simplest set of equations for a system of this kind simply states that each population has an equilibrium value which is a function of the existing values of all other populations. This gives us n equations and n unknowns immediately, and if this set of equations has a solution in positive values for each population, an equilibrium is at least possible. Whether it will actually be attained or not, of course, depends on the dynamics of the system, for the dynamic processes of the system may change these equations in the course of time. For instance, if any biological population in the course of the dynamics of the system falls to zero, it will never recover, and all the other equations will have to be changed, and the final equilibrium likewise will have to be changed. The system could also be formulated in dynamic terms by supposing that the rate of increase of any population, which would be negative, of course, in the case of a decrease, would also be a function of the size of all other populations. This gives us a set of simultaneous differential equations, which, again, may have a solution in terms of a path for all populations. This path may or may not move toward an equilibrium in which the rate of growth of all populations is zero. This is the model which clearly underlies the Walrasian equilibrium in economics and it is quite a small step to move from the special case of commodity equilibrium to the general case of ecological equilibrium. Economics may certainly, therefore, claim to have made a fundamental contribution in this regard.

Arising out of this interest of economists in general equilibrium, it is not surprising that it was an economist, Paul Samuelson, in his *Foundations of Economic Analysis,* who gave the first clear exposition of the principle that equilibrium of any system had to be derived from its dynamic path and that we could not find out about the stability of an equilibrium from an inspection of the equations of equilibrium alone. A stable equilibrium was itself a property of the dynamic path of a system, so that even the simplest properties of any equilibrium system depended in the last analysis on the dynamic process of which it was, in a sense, a special position. Equilibrium is simply a dynamic process in which the dynamic path of the system leads to a reproduction in successive states of some initial equilibrium state. "Staying the same" is simply a special case of "changing." Stability has to be seen as a subspecies of change.

Another area in which economics has made an important theoretical contribution to other disciplines is in the theory of behavior, where the great contribution of economics has been the theory of maximizing behavior, that is, the assumption that the behavior of an organization could be explained on the grounds that it was trying to maximize some internal variable. In the most general case, what is maximized is simply "utility" or an ordinal preference function. All we mean by this is that behavior consists of doing what the behaving organism or organization "thinks is best at the time." One difficulty with this theory is that it becomes too general and hence without much content, stating little more than that organizations and organisms do what they do simply because, if they didn't think that what they did was the best thing to do, they wouldn't do it.

The economist's concept of "revealed preference" is simply one way of finding a pattern in nonrandom behavior. In any behavior that is not random, we will be able to find some sort of revealed preference, that is, we can postulate a preference function from which we can deduce the behavior which is actually observed. If an amoeba, for instance, "chooses" a piece of food rather than an adjacent stone, we say it is because it has a preference function on which food ranks higher than nonfood. To say that when we have said this we have not said very much may be right. Nevertheless, even if we have not said very much, we have said something. This is at least one way of describing nonrandom behavior, and those who think it is merely empty should at least accept the challenge to describe it in some other way that seems more useful. The whole concept indeed of a cybernetic system with a detector-selector-effector mechanism implies that the selector has some principles according to which selection is made, which is precisely what we mean by a preference function.

I must still confess to some qualms about this formulation, mainly because there are clearly cases in which it seems anthropocentric, to say the least, to suppose that the selection process of the system involves preferences. In the general evolutionary theory of natural selection, for instance, it is only by stretching the language, perhaps beyond the point of legitimate strain, that we can say that the species which in fact actually survive are "preferred" by the selection process. If we had a satisfactory model of natural selection we might find that this would also throw a great deal of light on "artificial selection," that is, the phenomenon of choice and decision. The awful truth is, however, that we do not have any adequate models of natural selection. There is no general mathematical model of the evolutionary process, and it is by no means clear that this is even possible. We do have something like special patterns of selective processes, as, for instance, in the theories of Sewell Wright, but we do not really have anything that can be called a general model of the whole evolutionary process. Until we do, the economic models of selection by preference and choice have a great deal to recommend them and, provided that we recognize that the language is dangerously anthropomorphic, we can apply these principles even to organisms as simple, or complex, as the amoeba.

The economic model of behavior perhaps has had a greater impact on the applied psychological sciences of management science and strategic science than it has on psychology directly. The concept of optimization is quite fundamental to management science and to operations research, even though it is not always clear what is optimized. Extensions of this into Herbert Simon's "satisficing" are simply special cases of the optimizing principles, which states that any position of the organization in which the maximand, that is, the criteria of success, is below a certain level is regarded as unsatisfactory and any position above this is regarded as satisfactory. This is easily seen to be a special case of the general maximizing principle if we visualize the preference function as being more like a mesa with a flat top than it is a peak with a sharp maximum. In many cases a mesa view of the preference function seems to be realistic, that is, there is a considerable area of choice within which we are relatively indifferent, but beyond this area preference may fall off very fast in "cliffs." Game theory, likewise, can be seen as an extension of the simpler model of maximization, with cases in which the outcome of one person's choice depends on the choice of another person. It is certainly no accident that the classic work in game theory by Von Neumann and Morgenstern should have been entitled *The Theory of Games and Economic Behavior*. Here we see economics moving toward more general systems, though again almost at the cost of ceasing to be economics.

Theories of economic behavior have had a substantial impact on political science in the last generation, which has been moving more and more in the direction of political economy. A number of authors—Lindblom and Dahl, Anthony Downs, and Riker—have used what are essentially economic models in the interpretation of political behavior with considerable success.

A problem which emerges out of economics, but which has a highly general significance, is the problem of suboptimization, that is, under what circumstances does the attainment of some kind of optimum in part of a system preclude the attainment of an optimum for the whole? Welfare economics and the theory of perfect competition is one of the few areas of the social sciences which deal with this problem, yet it is a problem of great generality. Many of the failures of organizations, for instance, are a result of suboptimization, which could almost be defined as finding the best way of doing something which should not be done at all, or more generally, finding the best way of doing something particular without taking account of the costs which this solution imposes on other segments of the system.

Another field in which economics has made a substantial contribution toward a general system, at least in the social sciences, is in the field of international systems. I regard this indeed as one of my own major contributions. My book *Conflict and Defense* is an attempt essentially to apply a body of general theory, much of which comes out of economics, and especially out of the theory of oligopoly and the interaction of firms, to the problem of the interaction of states in the international system. It is indeed a general theory of

84

viability and survival, which comes out of economics and which has applications not only to other social systems but also perhaps in the biological field. It is closely related, for instance, to the theory of territoriality. It is relevant also to the theory of the niche and to the determinants of niches. Thus any organization in competition with others will find that its advantage in the interaction diminishes as it goes away from some kind of "home base," so that at some point the advantages of any further expansion fall to zero. This is what I have called the "boundary of equal advantage" between two organizations, but the concept could easily be generalized. It is these boundaries of equal advantage which really define the niches of an ecological system. Economics has made an important contribution here in location theory, especially in the work of Loesch, who demonstrated that, even if we start off with resources and population distributed uniformly in the geographical field, the sheer pressures of maximizing behavior will force the field into clusters and structures and will indeed create what are in effect niches in what previously had been a uniform field. This is a demonstration of great importance, which has received surprisingly little attention from the biologists.

When we look at the other side of the coin, the impact of general systems on economics, we find unfortunately that the record is meager. In the last generation certainly economics has pursued its own way, with very few influences from outside. The attempt of Parsons and Smelzer, for instance, to produce a sociological contribution to economics, seems to have had virtually no influence on the economics profession itself. The impact of psychology has also been confined to a few pioneering individuals, like George Katona at Michigan, and even here the main impact has been from the empirical rather than from the theoretical side. Even my own interest in these matters, I think, has been regarded as an amiable eccentricity by most of my fellow economists. This may be, as I suggested earlier, because general systems has made the greatest impact on those disciplines which felt the lack of a systematic theoretical core, which economics did not. Nevertheless it seems to me that this isolation of economics from one of the most interesting movements in thought in the last twenty or thirty years has been most unfortunate, and as a result economists have missed many opportunities for learning things which would have been useful to them even in their strictly professional capacities.

Thus one would expect that cybernetic theory, which has been so important in the development of general systems that some people have almost identified the two, I think quite falsely, would have had some impact on economics. Feedback mechanisms, for instance, are of crucial importance in a great many economic systems. If one is looking for an explanation of economic cycles, or fluctuations, either of particular speculative markets, such as the stock market, or in the economy in general, the feedback model is extremely useful. It is capable of explaining not only regular fluctuations, in the case of equilibrating (negative) feedback, but is also capable of explaining disequilibrium processes, as in the case of destabilizing (positive) feedback. Yet there has been astonishingly little use of this model. Perhaps as a result of this failure to take a significant intellectual tool simply because economists have not made it themselves, economists seem virtually to have lost interest in the theory of fluctuations, and what work has been done in this has been of an extremely mechanical nature, using, for instance, spectral analysis, which is an elegant way of detecting probably nonexistent cycles and throws no light on the real structures and processes which underlie fluctuations. There is very little appreciation among economists, for instance, that the Great Depression from 1929 to 1933 could be explained in very large measure by a destabilizing feedback process in which a decline in investment produced a decline in profits and that a decline in profits produced a further decline in investment, which again produced a further decline in profits, and so on, until by 1932 investment was almost zero and profits were negative. There has been some work of this kind on inventory cycles, which really is something like a feedback mechanism, but this is highly specialized and not widely used.

Another area in which general systems should have made a much larger impact on economics than it seems to have done is in the theory of the optimum size of the organization. One of the real triumphs of general systems theory, for which Von Bertalanffy must take a great deal of credit, is the demonstration that the processes which lead to the formation of organizations tend to exhibit "equifinality" in the rather special sense of self-limiting growth. This is a consequence of the principle of "allometry" which is based fundamentally on the thesis, which is really an identity, that as the growth of any particular structure proceeds, in the absence of change in the pattern of the structure itself, volumes will grow eight times as fast and areas will grow four times as fast as linear dimensions. This is why the whale has to live in the ocean, simply because, if it were a land animal, its legs would have to be bigger in cross section than the animal itself.

The same principles apply to social organizations, though the relationships here are not so simple. Here the main limiting factor is clearly the lines of communication, particularly up and down the hierarchy, which grow at a much slower rate than the total organization. Hence organizations eventually limit their own growth simply by the sheer difficulty in getting communications from the "surface" of the organization, where it is in contact with its environment, into the decision-makers who are not in direct contact with the environment, but have to make decisions in the light of increasingly less realistic images of the world. Acephalous, nonhierarchical organizations, like a democratic family or a commune, or even a producers' cooperative, have even sharper limits on scale, simply because the number of people who have to talk to each other increases much faster than the number of people

87

in the organization. Groups employing participatory democracy have the same tendency for fission as does the amoeba, for very much the same reason.

The economic significance of this principle is, of course, that the firm, like any other organization, has an optimum size and that this optimum depends in part, at any rate, on the internal diseconomies of scale setting in beyond a certain point, mainly due to the difficulty of maintaining communications in a large organization, even when it is hierarchical. Centralization produces failure to optimize because of the breakdown of the communication network. Decentralization on the other hand produces suboptimization, from the point of view of the organization as a whole, and likewise produces a failure to optimize. The cycles of centralization and decentralization that we see in nearly all large organizations in the Soviet Union, the Catholic Church, and General Electric, almost certainly arise out of this principle. If there is something wrong with every alternative, one tends to try a succession of wrong things in the hope that maybe one of them will turn out, which it never does. This principle of internal returns to scale is very important in explaining why some industries, like agriculture, have rather small firms and hence can maintain something like perfect competition, while other industries, like automobiles, have very large firms and hence produce either monopoly or oligopoly.

Another general system of great potential importance in economics is that of population analysis. Any process involving a set containing elements in which the date of entry into the set, or "birth," and therefore the age at any particular moment, of any element can be identified, and in which also the elements leaving the set, that is, "death," can be identified, deserves the name of "population." If there are functional relationships between the age structure of the population at a moment of time and the total number of births and deaths, the population can be projected into the future by a fairly simple system of what are essentially differential equations. Populations do not have to be biological; we can perfectly well consider the total number of automobiles, for instance, as a population, and the future population of automobiles can be projected in very much the same way that we project populations of human beings or of deer. I did this in fact in 1954–55.[1] This is indeed the key to a good deal of economics that goes by the name of capital theory, capital being simply the total population of valuable objects that exist at a moment of time. The "period of production," which has been a prominent concept of capital theory, is virtually the same thing as the expectation of life at birth of the elements in a population. The principle that in an equilibrium population the total population is equal to the annual number of births or deaths multiplied by the average expectation of life at birth appears in economics rather crudely as the capital-income ratio, which has been important in certain development models, though often used quite illegitimately because of the failure to recognize that disequilibrium populations rather than equilibrium populations were involved.

Finally, one might look at the failures of economics in the present generation and see how far these might have been due to the failure of economics to use insights from other disciplines, and especially of course from general systems. The great success of economics in this period from, say, the end of the second world war, has been mainly due to the capacity of the Keynesian system to suggest policies which at least prevented large-scale unemployment of the kind that we had in the 1930's. Even though the level of employment in the United States has not been by any means wholly satisfactory, hovering as it has in the last twenty-five years between about 3 percent and 6 percent, it is certainly much better than the 25 percent unemployment we had in 1932. I have often drawn the contrast between the twenty years that followed the first world war, from, say, 1919 to 1939, with the twenty years that followed the second world war, let us say, from 1945 to 1965. The first period was a disaster. The recovery of Europe from the first world war was very halting and even though the prosperity of the United States in the 1920's was quite real, the Great Depression was an unmitigated disaster, and the thirties slid almost inevitably into the second world war. By contrast, the twenty—now twenty-seven—years after the second world war were quite successful; there was no great depression, there was a substantial economic development, especially in the richer countries, and since 1961 at least we seem to have moved further from the "third world war" to the point even where there is quite a high probability that it will never happen at all.

Some of this success is due to the Keynesian economics, which operated very much within the traditional framework of economics itself and did not draw on either theoretical models or information from other disciplines. So far the economists might claim that economics was self-sufficient, and that its successes were striking enough so that it did not really need anything from other disciplines, least of all from the amorphous body of ideas called general systems. Nevertheless there are a couple of flies in this moderately sweet-smelling ointment. One is the very uneven success which economists have had with the advice that they have given to the poor countries seeking to develop. There are a few notable exceptions, but on the whole the tropical world has not done conspicuously well economically in the period since 1945, even though it has seen the virtual liquidation of the European empires, with the exception of the Portuguese. It seems quite reasonable to associate this relative failure of economics in the field of development with its inability to go beyond rather mechanical models, which abstract too much from the enormous complexity of real societies. It is simply not enough to use gross economic aggregates. The developmental process is essentially a process in human learning. It involves enormously subtle relationships of status, authority, threats, and persuasion, as well as exchange. The plain fact is that we have not yet produced an adequate and total model of the developmental process in any society, and especially in those tropical countries of

90

ancient and complex cultures, and our knowledge of the total processes of these societies is skimpy indeed.

A somewhat related problem has been the failure of economics to deal with the problem of policy toward deteriorating cultures and deteriorating cities, especially in the rich countries. The failure of economics here, I think, is due again to its obsession with rather simple models of exchange and its refusal to recognize that many of these problems involve one-way transfers, that is, a grants economy, which depends in turn on an enormously complicated human learning process in matters of status, identity, community, benevolence and malevolence, and so on. The very fact that there is an Association for the Study of the Grants Economy suggests that conventional economics has failed in this regard, again perhaps because it was too shut up within its own particular system and frame of reference.

Perhaps the most visible and spectacular failure of economics in this period has been its failure to deal with the problem of how to get full employment without inflation. This is a failure of the Keynesian model itself. In the United States, for instance, we have had almost continuous inflation, at least as measured by price indices and national income deflators, since about 1939, and there are no signs of our being able to control this. President Nixon's wage-price policy of August 1971 indeed is a little reminiscent of King Canute ordering the tide to go back, and, while it may have a temporary psychological effect, it is extremely unlikely that it will provide a solution to the problem. The reason for the failure of economics, here also, perhaps is that it has lived too much within its own framework and has neglected to study the larger system of price determination with models that might have been derived, for instance, from epidemiology theory, or from the theory of the spread of fashion, or from the theory of communication networks, none of which has been at all familiar to economists. I suspect indeed that we are not going to solve this problem until we are able to identify the leading communication chains in the structure by which the inputs of information, which lead to changes in prices and wages, actually function.

It is easy to give economists good advice; it is very hard to make them take it. Nevertheless the defects in economics at the moment are so glaring that one hopes that a new generation will develop, with an appreciation of what can be learned in economic models from other sciences and from general systems, so that the general systems economist will not always remain as a voice crying in the wilderness.

6

Communication Systems

LEE THAYER

I T would be easy to agree with Von Bertalanffy that, as he observed in one of his recent papers, there are questions which are far from satisfactorily answered about what general systems-theoretic statements can be made regarding "material systems, informational systems, conceptual systems, etc."[1] The same observation could certainly be made with respect to communication systems. In spite of much talk to the contrary, there has been very little exploitation of the potential of general systems theory in the study of communication systems and, indeed, very little systematic interest among general systems theorists and advocates in the study of human communication systems.

There are reasons why this is so, reasons which are pertinent to the future relationship between general systems theory and human communication theory. In this paper, it will be my purpose first to examine those reasons, and some of their implications; second, to discuss the peculiar conceptual difficulties involved in accommodating general systems theory to the study of human communication, and vice versa; and third, to propose a framework for mutual development in the future.

Terminological Difficulties

The most obvious reason for the lack of a growing synthesis between general systems theory and human communication theory is terminological. Von Bertalanffy himself has often seemed to equate human communication theory with information theory or electronic communication theory,[2] both based in the physical properties of the materials, devices, and particular medium (air, water, space) involved in the electronic transmission and acquisition of data. Many spokesmen for general systems theory do not address themselves to the phenomenon of *human* communication at all.[3] Among those who do,

95

many seem to assume that there are no theoretically significant differences between an electronic data system and a human communication system.[4]

This terminological confusion, both real and potential, is perhaps even more pervasive among those who identify themselves as human communication scholars or researchers. It is quite apparent, for example, that the mechanical, linear model of communication which has been widely adapted from the original Shannon and Weaver work (and later from Wiener) is the most common model of the communication process in use by researchers of human communication (that is, $A \longrightarrow M \longrightarrow B = X$, where the sender, A, "communicates" some message, M to the receiver, B, with X result—with or without the added complications of "form," "medium," "noise," etc., which are but refinements on the basic model). This model for the study of communication also presents some fundamental conceptual difficulties, to which I shall return below. A term does not explain a phenomenon, so the terminological difficulties that follow from importing (or analogizing) the "communication" of electronic signaling systems into the study of human communication are considerable.

A second, more mundane, and perhaps therefore less obvious terminological difficulty inheres in the fact that a term which is in popular use has a richness of meaning (and therefore an ambiguity) that may limit its theoretical usefulness. The easier it is for a layman to use the term "system" in his everyday parlance, the less likely he is to concern himself with its theoretical (i.e., within the discipline) meaning. This has long impeded substantial development of theory in human communication. Everyone "knows" what communication is. We all do it, and more and more of us can talk casually about it. As Whitehead suggested, the more familiar something is to us, the more difficult it is for us to subject it to systematic, or scientific, inquiry.

The commonplace notion of human communication, which is that there is some transmission or sharing or exchange of "information" among people is, unfortunately, widely assumed by scholars in other disciplines. While there is probably less of a tendency among those who concern themselves with general systems theory to assume away the issues involved in this cavalier fashion, there is perhaps too much, and these faulty assumptions* contribute to the terminological difficulties between the two fields.

Finally, it should be noted that Von Bertalanffy has on at least three occasions discussed at some length a general theory of symbolism.[5] What he has said and written about symbolism could hardly be faulted; it is understanding and insightful. Yet the terminological difficulty these treatments present lies in the implication (surely not intended) that a general theory of symbolism would suffice as a theory of human communication.[6] To say that man is an *animal symbolicum* is certainly one way of distinguishing man biologically from other animals. But it explains very little. Man's symbol-using capacities are *means*; the human and social *ends* to which man thereby propels himself are not given in the mere fact of this unique capacity— as, indeed, Von Bertalanffy has recognized in his concept of human symbols as "freely-created."

The difficulty here is that symbol-making and symbol-using[7] are but one limited aspect of the process of human communication. We must avoid the temptation to confuse "symbol-using" with "communication." (Nor is a theory of language a theory of human communication.)

All such difficulties are, of course, more than just terminological. They are philosophical and conceptual as well.

Philosophical Difficulties

There can be little doubt that early writers like Mead and Cooley in social psychology, Bernard and Von Bertalanffy in biology, Vico and Bentley and others in social theory, and Köhler and Lewin and Goldstein in psychology, among many

---

* Just why these assumptions are "faulty" I will explain below.

others, have contributed greatly to the gradual shift from physicalistic to "organismic" thinking in the life and behavior sciences. And there can be little question that more contemporary writers like Boulding, Rapoport, Buckley, Laszlo, Vickers, Easton, and the several authors represented in the Gray- and Rizzo-edited volumes, *Unity through Diversity: Festschrift in Honor of Ludwig von Bertalanffy*, are doing much to further the posture and the philosophy of a general systems theory in an increasingly wide range of scientific and applied disciplines.[8] Von Bertalanffy did indeed anticipate a very general trend away from the classical paradigm of science toward the more "holistic" or "systems" paradigms of modern thought, from biology[9] to ecology to space engineering.

Yet any new paradigm, particularly perhaps a popular one, does not obviate the philosophical (or metascientific) bias which it brings to bear.

I do not wish to be misunderstood here. Von Bertalanffy was careful to note and to answer those whose skepticism would be based in fear of dehumanization at the hands of the systems "scientist." It would be silly to argue that any theory which is enlightening is intrinsically either good or bad for humanity.

What I do wish to point to, as the main philosophical difficulty, has nothing to do with this kind of misplaced skepticism. It has to do with the assumption that the processes one encounters from the biological through the sociopolitical are continuous. Or, put in another way, it has to do with the assumption that a general theory which holds for lower-order phenomena will therefore necessarily hold for higher-order phenomena. Or, put still differently, it has to do with the point at which a theory, in order to be generalized, is overly analogized.

The source of this difficulty lies in a whole set of deeply embedded tacit assumptions which are the warp and the woof of the philosophy of science litself. To discuss them at any length

98

would go far beyond the scope of this paper. Two examples must suffice.

There is first the fact that the boundary between the so-called "hard" sciences and the so-called "soft" sciences and humanities can be freely traversed in only one direction. We intuitively feel it to be legitimate that the expertise of scientists and engineers be brought to bear on all kinds of human and social issues. (It is presently federal policy to rechannel unemployed scientists and engineers into the solution of social problems.) But it would seem much less likely to us that a poet or a social psychologist would have anything to contribute to the leading edges of genetic or atomic theory. This kind of variable impedance to communication across that "boundary" suggests not a technical or scientific, but a philosophical, difference. And I sense that there is some philosophical difficulty of this sort which impedes the mutual interpenetration of general systems theory and human communication theory.

Second, Von Bertalanffy has himself argued that "models in ordinary language . . . have their place in systems theory."[10] But I think he also implies that such a condition should be looked upon as but a temporary shortcoming and that eventually a theory (in sociology, for example) can be "scientific" only when it has been formulated mathematically. The point to be raised is not whether this is "right" or "wrong." It is whether there may be some philosophic or metascientific question about the modeling of a process of which the model-maker is himself a model. Outside of fields such as astronomy, science can predict, in the traditional sense, only that which it has the technology to control.[11] Since what is specifically human about man is continuously being modeled (created) by man (himself, his cultures and societies), can there be any model of man other than man?

The difficulty here, in short, is that much of what has been written and said of general systems theory to date is heavily, though tacitly, freighted with all kinds of philosophic residuals and transforms of physicalism (e.g., the addition of "feed-back" doesn't basically alter the S–R conception of human communication which is implicit in the conventional model), and this has impeded the mutual development of general systems theory as well as human communication theory in the study of human communication systems. This is, especially, no criticism of general systems theory. As a "point of view" (Boulding) or as an "approach" (Monane, Churchman[12]), its potential awaits exploitation in this relatively neglected area (of human communication theory).

The aim of general systems theory to bring unity to the sciences through the discovery or exploitation of theoretical isomorphisms is laudable, and much provocative and stimulating work has followed Von Bertalanffy's statement of this philosophy of general systems theory. As always, there will be those who in their zeal would sacrifice empirical discontinuity for theoretical continuity. There may be such significant empirical differences between biological systems and human communication systems that different theories are called for. If there are, we ought not let a philosophy of unification of knowledge stand in our way.

These terminological and philosophical difficulties are not independent of the conceptual difficulties one faces in approaching the study of communication systems from either the point of view of general systems theory or of human communication theory.

Some Conceptual Difficulties[13]

The major conceptual difficulty remains that of the faulty and inadequate model of human communication employed—implicitly if not explicitly—by the majority of general systems theorists and communication theorists alike.[14] That model, $A \rightarrow M \rightarrow B = X$, often gives the illusion of explanation, where it is at best but primitively descriptive, and at worst obscures its own weakness.

This model comes, of course, from electronic communication theory, but the analogy to human communication does not hold, as will be explained in more detail below. It may be sufficient here to note only that a coded pattern of data in electronic communication theory, which the transmitter has been designed and programmed to send and the receiver has been designed and programmed to receive (and sometimes to process in some way)—that such a coded pattern of data, the "message," must be invariant for the transmitter and the receiver under all conditions. This is the whole point of reliable telecommunication. For humans, on the other hand, the "message" (which itself is a misnomer) must vary with the context. The sergeant who got the desired response from his order "Attention!" in boot camp would have more than some little difficulty in getting the same response from the same men on the main streets of their own home towns after they had been mustered out—in spite of the fact that the "message" is employed by the "same" sender and the "same" receivers. If my neighbor said, "I intend to take 25 percent of your income for my own use," we would think he was out of his mind. The "same" message from the appropriate government agency is certainly a different message, even though the coded pattern of data could be precisely the same.

Closely related is the conceptual confusion that has resulted from analogizing the concept of "information" from electronic communication theory, information theory, and cybernetics into the study of human communication. The absence of input to a receiver in a signaling or control system is just that. Silence from the other end of a telephone conversation between two humans could be the most profound message of all. When a human says, "I'm not talking to you," that may be precisely what he is doing; a computer address, to the contrary, either works or it doesn't. A radio receiver has to accommodate the incoming "information" for which it is designed, like it or not; humans can "tune out" or even "turn off"! "Information overload" in the information theory sense is a matter of quantity. Between husband and wife, it may be sheerly a matter of timing: there are those times when a single word may be too much.

Such examples could be multiplied ad infinitum. What they reveal is that the "information" of information theory, which has been largely adopted as an adequate concept for purposes of general systems theory, is not the "information" of human communication. The two phenomena are qualitatively different, and they must be kept conceptually distinct.

At the other end of the scale—and this largely from the concern of some humanists with human communication—there is the tradition of assuming the central issue to be one of "understanding" or of "meaning." From this point of view, the end of all human communication is understanding. The study of human communication has more than once in man's intellectual history been reduced to the study of meaning. This approach is equally misleading. The meaning of "Crime doesn't pay" depends upon whether one is a criminal or not. The fact that many Chinese are perplexed by the fact that Americans refer to their form of government as a democracy is more than just semantic. The assumption made by the "content analysts," that they can extract the meaning of a document or an utterance, is the third-person fallacy in spades.

We realize that meaning resides not in messages but in people, but we seem to have continuing difficulty accommodating this fact in our models and theories of the human communication process. The notion that words or statements "refer to" something in the "real" world is the most naïve and primitive concept of human communication there is, yet in some quarters it is still the guiding paradigm.

There are many other such conceptual difficulties which have impeded the mutual interpenetration of general systems theory and human communication theory. For these, the interested reader may consult the several references suggested. It is time now to turn to the development of a conceptual framework for apprehending human communication from a systems-theoretic point of view.

Toward a Conceptual Framework

Two of the most troublesome metatheoretical stumbling blocks in the life and behavioral sciences—and for reasons peculiar to those areas which concern themselves with living systems—have been those of the *unit* and the *level* of analysis. In a universe of nested systems of human thought and action, which is the most advantageous theoretical boundarying—horizontally and vertically—of the phenomenon one wishes to study? Nowhere is this issue more central, or the consequences of poor choices more detrimental, than in human communication theory. Is the basic phenomenal unit to be that of the message, of the message plus the sender, . . . plus the receiver, . . . plus the response, . . . plus the consequences, . . . plus the context, . . . or what? All have been used as the basic unit of analysis, more often by default than not. The consequences have been a muddled and nearly equivocal hodgepodge of research and, to this point at least, a failure to settle upon a theoretical foundation which is sound enough to be built upon. (The traditional $A \rightarrow M \rightarrow B = X$ model, as has been pointed out, is neither an adequate nor an empirically sound enough foundation upon which to build.)

Although one would logically select a different *level* of analysis depending upon his particular objectives—the intrapersonal level, the interpersonal, the small group, the large enterprise, the enterprise $\longleftrightarrow$ environment, the technological, etc.—a theory must evolve from some fixed reference point, the basic unit of analysis.

As I hope to show, for human communication theory, this basic unit should be the communication system. But first some preliminary definitions.

(i)   Preliminary Definitions

The very limited conception of communication, as something that one human *does to* another human, is the point of departure. This is a beautiful example of the obsolete kind of physicalistic, linear, two-body type of thinking which Berta-lanffian general systems theory so well reveals and denies. What seems to be clearly unavoidable when one tries to recon-ceptualize these phenomena in systems terms is that communi-cation is one of the two basic life processes—the one being the acquisition and processing of energy, the other the acquisi-tion and processing of information. What one sees in looking at the increasing complexity of living systems from the single cell to the human is that the basic process—some living sys-tem taking-something-into-account to some end—obtains at all levels, and that what we observe at the human level is in one sense no more than a highly sophisticated form of what occurs at all levels.

Indeed, what distinguishes human communication is not that people talk to each other, for one can observe this phenomenon among all the mobile and socially specialized creatures of the earth. What distinguishes human communica-tion is that people talk to themselves. If there is a discontinuity, it is here—that humans have the biological capacity for self-reflexivity, a capacity whose exploitation seems to parallel the level of civilization achieved.

From a communication point of view, this peculiarity of man has a very profound implication. Because a man can con-ceive of himself as a part of the process of conceiving of his outer world he, unlike his fellow communicating creatures, is not bound biologically (e.g., by instinct) to any *necessary* relationship between what is and how he relates himself to what is. The scout bee which dances out the location of the pollen is biologically constrained to provide a message which has a necessary relationship to what has occurred. A man can tell it any way he wants to, in dance, song, chart, computer program; he can even tell a lie about the direction of the food he has found.

There are a great many ramifications—for human communi-cation—of this peculiar capacity of man's, and some of these have been identified by Von Bertalanffy under his concept of

104

"freely created" symbols. Among others, there is the well-known constraint on the "lower" animals of presentness; a dog can be trained to alert you to a present intruder, but he cannot be trained to tell you that there was one yesterday or, more crucial, that he is expecting one tomorrow.

The ubiquitous process in all communication is therefore the taking-into-account of something by some living system to some end. This is of the order of a first principle, and is therefore the foundation upon which all ensuing theory must be built.

One further and important implication of man's peculiar capacity for self-reflexivity in the communication process deserves attention here. Because there is some necessary relationship between lower-order creatures and their environments communicatively, one can justifiably speak of *teleology*: such living systems are ever—as long as living—in the process of becoming what they *are*. But we need another term for what man *can* do, even if we have no more evidence for the capacity than what a few men have achieved. Elsewhere I have used the term "telesitic" behavior,[15] to distinguish it from "teleological" behavior. The term is not entirely satisfactory, but we need some concept to account for the fact that men, unlike their fellow creatures, can undertake to become (socially, psychologically, etc.) what they *are not*. A man is ever in process of becoming what he is; but he may also elect to engage in the process of becoming what he is not (presently). The fact that man is telesitic as well as teleological has a great many fundamental ramifications for understanding his communication behavior.

Man's consciousness (the by-product of his self-reflexivity) frees him of any necessary communicative relationship with his environment. A further basic ramification of this peculiarity is this: all other creatures enter the world essentially *pre-formed* to take-into-account that which they can and must. Not so, man. He must be *in-formed*, and this, by his fellows. What is therefore real, important, relevant, etc., for man must

105

be learned by him (or "taught" to him). A common cricket (Gryllidae) is merely *informed* by certain signals which he has been pre-formed to receive and respond to. A man is both informed and *in-formed*. Because he has received and acted upon it once, no "message" will ever be exactly the same to him again. A man evolves as a human out of his communication experiences. He is irreversibly altered.

In short, what is "reality" for man's fellow creatures is biologically provided; what is conscious reality for man is created by him—for better or for worse, a question that cannot be answered either by man's genes or his biological complexity.

(ii) Communicational Realities

One of the distinctive features of human communication and intercommunication, as indicated above, is that there is no necessary relationship between the environment and the way it is consciously construed by men (beyond the purely physiological conditions of life; yet, as we have witnessed in this century, even these may be near-tragically misconstrued). Certainly there is no necessary relationship between men imposed upon them by blind nature. To the social and political and economic forms of relationship by which men live, there would seem to be no end.

What this implies for civilized man is that the conscious realities by which he lives are largely communicational realities—that is, those things and ideas and goals and means which have "reality" only in the sense that they can be and are talked about in certain ways. Beyond what an individual man can test with his own physical senses lies that whole universe of realities and unrealities (the "noosphere") by which people live, by which they become conscious of themselves, which make possible their "telesitic" behavior. "Love" and "nation" and "money" and "man's destiny" are good examples. They are more than abstractions, more than symbols having some concrete referent somewhere; they are communicational realities, having only that reality which is provided by the fact and

106

the way in which two or more people talk about them. "It takes two of us to create a truth," says Kahlil Gibran, "one to utter it and one to understand it."

The most basic and profound communicational reality of all is "I." Each of us uses it when we talk to each other. But to what, precisely, does it refer—to the physical me, to the psychological me that I conceive of, to the psychological me that *you* conceive of? To what, indeed, can it refer? There is no precise referent. It is a grand metaphor, a metaphor which we humans find indispensable when we want to talk to each other, about each other.

The criterion by which communicational realities are to be assessed is therefore not their isomorphy or veridicality with what *is* (i.e., with what exists, with what is "real"), but that they "work" (i.e., that they do or do not enable people to achieve their ends vis-à-vis one another, or vis-à-vis their environments). This little aphorism, I believe, captures something of this profound difference between human and nonhuman communication, and of the existential reality of communicational realities:

> For all other creatures, communicatively,
> Whatever is, works.
> For man, communicatively,
> Whatever works, is.

(iii) The Functions of Communication and Intercommunication

In all of those organisms below man on the phylogenetic scale, then, the function of communication is that of *implementing* the by-*nature*-designed ordering of the individual into its environment. The function of intercommunication[16] is that of *implementing* the by-*nature*-designed ordering of individual organisms to each other.

For man, the self-reflexively conscious creature, the function of communication is that of *enabling* the by-*man*-designed ordering of the individual into his environment, and the function of intercommunication that of *enabling* the by-*man*-designed ordering of individual men to each other.

The distinctions are crucial. Non-self-reflexive creatures are *pre-formed* to take-into-account their environments and each other in certain ways that, we assume, somehow serve the ends of nature. Men are, through their uniquely long and dependent periods of socialization, *in-formed* by other men to take-into-account their environments and each other in ways that, we assume, somehow serve the ends of man.

The conditions for "successful" intercommunication give us further insight into the nature of both human and nonhuman communication. The *necessary* condition is of course that one individual signal or say something (or not) to one other (or more) individual(s). The *sufficient* condition is not only, however, that the other individual has been properly in-formed to comprehend what is said, but that he can accept the statement as having validity or legitimacy (i.e., that it can be trusted to be acted upon in terms of the relationship between the individuals involved). For all of the essentially pre-formed animals and insects, the "signal," so ethologists tell us (as in the "Kumpan theory" of nonhuman intercommunication), is both the necessary and the sufficient condition for the interlocked response. But what the caveman needs to know when he is invited by one of his fellows to cooperate in a hunt for a bear for food is not only what the speaker "means" by the words or the pictographs he uses, but that he can accept what the speaker says, that he will not himself be slaughtered for food, that he will get his fair share of the bounty, and so on.

Human intercommunication is therefore dependent upon a special kind of metacommunicative "trust" (for want of a better term), which is not at issue for those creatures which are preformed for intercommunication.

Because there is no *necessary* relationship—for man—between what *is*, and what is said about what *is*, men may indeed lie to, deceive, dissemble, and distract each other. Nonhuman intercommunication is by nature's design largely *tactical*. The intercommunication of men is, by the design of those who practice it, often *strategic*.

Whatever the failings or the errors may be in either the case of creatures which are pre-formed or in the case of man, who is *in-formed* for communication and intercommunication, we cannot overlook the underlying functions of communication and intercommunication. These functions are ultimately— however diverse the species- or tribe-specific hardware and software for doing so—the *ordering* of the individual into his environment and the *ordering* of individual members to each other to some "social" or aggregate end.

The indivisible unit of analysis, then, is the organism together with that-which-is-being-taken-into-account—i.e., the communication *system*. At the communication level, it is the individual organism together with that part of its environment which it is presently taking-into-account in some way, and to some end. At the level of intercommunication, it is two or more individuals who are presently taking-each-other-into-account in some way, and to some end.

(iv) The Technologies of Communication and Intercommunication

The technologies of communication and intercommunication are often confused with the processes involved, and this should be briefly clarified here.

The technology of communication includes all of those organs and organ-amplifying devices which are utilized by individual organisms for taking-into-account their environments. For man, this would include all of his senses, his central nervous system, and his brain, and such devices as eyeglasses, radar, computers, television, and mental *schemata*, all of which amplify his own organs either in the acquisition or in the processing of environmental data.

The technology of *inter*communication includes all of those organs, organ-amplifying devices, and codes which are utilized by two or more organisms for taking-each-other-into-account. For man, this includes all of those communication technologies previously noted, plus all of those organs (as of speech) and devices (megaphones, radios, etc.) which are utilized for amplifying the transmission or acquisition of data about one another, plus all of the codes (languages) which are employed for doing so—whether verbal, kinetic, symbolic, etc. What is most important to note here is that man's languages—verbal and nonverbal—are a part of the technology of intercommunication, and should not therefore be equated with the process itself.

(v)   Communication Systems[17]

As defined above, the smallest logically indivisible unit of analysis for the systematic and scientific study of communication and intercommunication is the communication system. A communication system is comprised of the organism or the individual, together with that which is presently being taken-into-account, whether some aspect of its environment or another individual or organism.

There are two distinguishing characteristics of communication systems (particularly of inter-communication systems) *qua* systems which I believe present some difficulty for general systems theory.

First, there is the fact that in (inter)communication systems the components are *interdependent* in a rather special sense. In physical systems, and in "information" (read "data") systems, one logically assumes the identity of the components: i.e., one logically assumes that the same component would behave in the same way in any other system having the same characteristics. The chemist can safely assume that the 50 cc of purified hydrochloric acid he takes from vial A is just the same (assuming a reputable source) as the 50 cc of the same chemical which he takes from vial B, and that any compound he produces will be equivalent if he uses accurate measures and the same compounding sequence. For the "information" engineer, a "bit" of information is a "bit," regardless of the time of day or the number of times it is used.

Not so the components of human communication systems. An individual is not the same in one communication system as he is in another. He is not the same "element" in conversation with his secretary as he is in conversation with his wife or his children. There is always something which is significantly unique about an individual in every communication system in which he participates. The components of a human communication system are interdependent in that system in the sense that the only way they can be defined in *that* communication system is in relation to each other. No way of defining them independently of each other can fully explain their behavior in *that* system. Psychology and sociology, trying to emulate the posture of the physicist, have too long ignored this fundamental fact. General systems theory helps to point it up, but a human communication system is not the same as an "information" system. In an "information" (read "data") system, the data are the message. In human communication, "information" is the product of a human transform of the data. In human communication systems, "information" is what people do to the data.

The second distinguishing characteristic of human communication systems *qua* systems is that they are, in Vickers' term, *historical*.[18] That is, not only are the components of a particular human communication system never the same in other communication systems; they are never again the same in the "same" communication system. This is not merely a matter to be accounted for by some stochastic process. People who talk to each other, particularly when they touch upon vital or fundamental communicational realities, change each other, sometimes peripherally, sometimes centrally, but change nonetheless. The "information" engineer can assume that a "bit" can be used over and over again, retaining its initial identity. To the chemist, hydrogen atoms can often be bonded or compounded and then reclaimed, retaining their universal identity.

111

The products and byproducts of human intercommunication are irreversible. Whatever one person says to another can never be said (or comprehended) again in exactly the same way by those persons, not because the words or the "message" have changed, but because the persons involved (and/or their relationship) have changed.

These two distinguishing characteristics of human (inter-) communication systems present particular difficulties, I believe, not only to the communication theorist or researcher, but to the general systems theorist.

It is *in* communication (systems) with each other that individuals create, alter, exploit, or maintain those communicational realities upon which their relationship is ordered. Telecommunications systems (and their extensions) describe the physical or structural ordering of intercommunication; communication systems, by contrast, describe the *functional* ordering of people in a society.

Communication systems which are extended in space or time are therefore also, in Holzner's happy phrase, "epistemic communities."[19] Those who have been similarly *in-formed*, who have similar frames of reference, similar epistemologies, similar "reality-testing" procedures, and who thus mutually validate each other's ways of knowing, can be said to be members of the same epistemic communities. The so-called "youth culture" of today, which spreads throughout the world, is a good example. The scientific community is another. Executives in any culture seem to think more alike than do executives and workers in any given culture; they share an epistemic community.

The more complex the civilization, the less validate-able is any particular statement by any particular individual. We come to validate (or to assess as legitimate or illegitimate, "true" or "untrue," etc.) the onrushing torrents of statements we hear in a society such as our own, not on the basis of the statement, but on the basis of the statement-maker. Those who can be recognized as legitimate members of one's own epistemic com-

112

munities can be "believed" (i.e., can be looked upon as providing legitimate or valid or at least reasonably "true" statements, which can not only be understood, but can be acted upon with some degree of confidence as to the consequences).

One function of such extended communication systems is therefore that of *normalizing* access and interaction, thereby providing not only *a priori* legitimation of what is or is not said, but the protocols[20] for who can speak to whom about what, etc. What people say to each other orders their relationships; their relationships order what they can say to each other. This is the primary social function of all human (inter)communication systems. The logical and theoretical difficulty is that each is the determinant of the other.

Another major function of human (inter)communication systems is that they serve, in their extended forms, as "relevanting" filters on the world. If there were such a thing as a nonsocial man, he would have to determine for himself what of his total environment is relevant to him—undoubtedly each time he encountered it or wanted to accomplish some act vis-à-vis some aspect of his environment. For man in modern, highly complex, and highly civilized societies, what-is-going-on (WIGO) may today be in part provided by the so-called mass media, but what-it-means-to-humans (WIMTH)[21] is still determined *in* communication with others of one's own epistemic communities (i.e., in communication systems).

One further function of human (inter)communication systems that should perhaps be mentioned here is the integrating-reality-process, which has its source in individual humans, and social-process, which exists as epiphenomena of reality-process. What is at stake, always, are the conscious existences of individual human beings, which are anchored in the vital communicational realities by which they identify themselves and their relationships with others, by which they interpret what is going on in the world, by which they assess the value or the worth or the "truth" of all that they may apprehend or imagine. What gives social process its persistence, its stability, its equilibrium (as well as the opposites of these) is the fact that people are ordered to each other in terms of these communicational realities, and the fact that they may individually have continuous existences only so long as those communicational realities by which they have ordered themselves are either continuously confirmed, or legitimately and adequately altered or replaced. Whatever obtains, at the level of communicational realities, must evidence itself in the social processes involving those who hold, or are affected by, those communicational realities. Whatever occurs at the social level must ultimately be interpreted in terms of this reality process. People do what they can, what they must, and, occasionally, what they would. It is out of the communicational realities upon which human behavior is based that the fabric of social process is woven.

The two levels are integrated or "synchronized" in human (inter)communication systems. This dynamic and dialectic process is what accounts for both persistence and change in social systems (a criterion for which Parsons, it is well known, has argued).

(vi) Systems Within Systems: The Nested Hierarchy of Human Communication Systems

It could hardly be overlooked that, in fact as well as by definition, in a complex human society there is no such thing as an isolated communication system. A small, primitive society could perhaps be viewed as a single, extended communication system. Everyone talks to everyone else, and the new members are homogeneously *in-formed*.

In complex societies, there are multiple communication systems and sets of communication systems (hence the term "pluralistic") which overlap and interpenetrate each other through common or linking members or participants. In addition, every conversation involves individuals who are members of the same and of different higher-order communication systems, which are themselves nested in spatially or temporally larger, more comprehending communication systems, and so

114

on, from the most casual conversation to the most embracing and long-lasting human "culture" (which is itself a kind of extended communication system within which certain communicational realities are confirmed and perpetuated).

Paradoxically, however, it is just the complex or pluralistic society, in which a multitude of communication systems are hierarchically "nested," which provides the greater opportunity for the "mutation" of emergent epistemic communities through *excommunication.*

The members of different epistemic communities can identify themselves in three ways: through *independent identification* (because of lack of awareness of, or indifference toward, other epistemic communities, like the early European cultures, separated as they were by effective natural barriers); through *mutual identification* (i.e., by some tacit ordering of "roles" to some collaborative end such as family life, "community welfare," "corporate profit," and so on); or through *disidentification* (wherein we establish our unique identities by not believing or being or doing as the members of other communities or subcultures believe or do or are). It is the latter means of creating epistemic communities which requires excommunication: in order to avoid the possibility of becoming like you are, I must avoid intercommunication with you; there must be an effective barrier that precludes the mutual contamination of our mutually exclusive communicational realities, and this can be achieved only through excommunication.

In terms of the theoretical isomorphy which general systems theory seeks, it is not without some significance that this is precisely the same process—excommunication—which some biologists and geneticists believe accounts for biological mutation.[22] The whole process is not unrelated to what biologists refer to as "form-creating variability." *Within* given epistemic communities or communication systems, it is, as Von Bertalanffy has suggested, the "nondiscursive" mode of intercommunication which creates new existential possibilities, or variability or diversity, as well as the kind of perturbations which lead to revitalization of living systems, a subject to which I shall return below.

(vii) "Closed" vs. "Open" Communication Systems

In terms of susceptibility to perturbation from the outside, human (inter)communication systems may be viewed as ranging from relatively closed to relatively open. A relatively closed communication system is one which either dispermits outside participation once formed, or which permits only nonvital participation (i.e., participation in ways which can neither create nor alter nor exploit nor vitally confirm those communicational realities upon which the system is built). A relatively open communication system is one whose boundaries are relatively permeable, and/or one which is open to participation by outsiders in ways which could create, alter, exploit, or vitally confirm central communicational realities.

A marriage in much of the Western world is a fairly good example of a relatively closed communication system. At the more extended level, the Amish communities of south-central Iowa are examples of relatively closed communication systems. They educate their own young in their own ways, are economically, spiritually, and politically self-sufficient, and forbid the reentry of any member who has once been contaminated by the "outside" world.

Examples of relatively open communication systems in the Western world are more difficult to come by. Public school education in the United States, as compared to the licensed medical community, exhibits a significantly greater degree of openness; often the most uninformed opinion carries the greatest weight in bringing about changes. Some of the more successful so-called "hippie communes" are, at least in principle, an attempt to create a more open communication system; but there is some evidence that the communicational realities upon which they are based are more inviolable than are the mores of the larger "establishment." At the societal level, one might compare the open society of the United States and other

116

Western nations with Castro's Cuba, Mao's China, Stalin's Russia, etc.

(viii) "High" Organization vs. "Good" Organization

A relatively closed communication system is likely to suffer from overorganization, to evidence "high" rather than "good" organization. But it does not follow that a relatively open communication system will exhibit "good" organization.

An illicit love affair, which is perhaps an even more closed communication system than a legitimized marriage, is likely, perhaps because of the persistent threat from the outside, to maintain what might be called "good" organization. But once legitimized by marriage, the communication system is likely to become overorganized—bureaucratized, routinized, etc.

The communicative ordering of humans into their environments and to one another presents a human hazard at both extremes. Too much order, too much fixity, too much expectedness, leads to human devitalization. Too little order, too much fluidity, too little reliability has, eventually, the same consequence for humans. Imbalances often give birth to new relationships, new epistemic communities.

There seems to be an optimum level of organization for every communication system, which depends upon some level of internal (and, where appropriate, external) perturbation. A man who has never had occasion to question his communicational realities (his in-formation) will never reach his full human stature. A man who must constantly question them consumes himself (and often others) in the process.[23]

When people talk to each other in a relatively highly closed communication system, not only do they understand each other but they provide mutual validation of their statements by being who they are (i.e., by being members-in-good-standing of the same communication system or epistemic community). They may well "founder on the rocks of agreement" (Shands). The viability of communication systems *qua* systems would seem to depend upon some optimum level of "form-creating varia-

LEE THAYER

bility," lest it become discontinuous (out-of-sync) with all of the other evolving communication systems within which it is nested.

Yet it is some optimum degree of excommunication or discontinuity between and among the communication subsystems of the whole (e.g., society) that enables the variability and evolution-producing diversity which is necessary to the health and the viability of the larger whole.

Much work remains to be done in understanding this central paradox of the parts and the whole of the nested hierarchy of communication systems that is today's complex society.

(ix) "Effectiveness" vs. System Criteria

Such system criteria have rarely been focal to the perspectives and research of those involved in the study of communication. The guiding criterion for most human communication theory and research right up to the present has, unfortunately, been that of "effectiveness." Was the "message" effective? How can we make it more "effective"?

To see more clearly the extremely limited usefulness of this criterion, we need to consider again how it is that any message can be "effective." There are three basic origins of the effectiveness of any (inter)communication encounter:[24]

1. If there is to be a viable society of men, there must be at least one epistemic community. If there is to be an epistemic community, all new members must be properly socialized to participate. From the point of view of communication, what this means is that members must be enabled—*in-formed*—to "produce" certain kinds of messages, and they must be enabled—in-formed—to "consume" certain kinds of messages —at certain times, under certain conditions involving certain other members, and so on. To look at two homogeneous members of a homogeneous epistemic community and to be interested in the fact that they can "communicate effectively" with one another on certain topics is to miss what is at issue almost

118

entirely. They can do so because they have been "programmed" to be able to do so, and both to give and to receive mutual validation for doing so. Homogeneous socialization is one source of the "effectiveness" of human communication, and we should look to the source and not to the "message" to see what is going on.

2. Because most human intercommunication codes (languages) have a grammar and are therefore generative, members even of different epistemic communities may talk to each other effectively—if they want to, and if they can agree upon a superordinate scheme for message validation, one which neither adds to nor subtracts from the separate schemes for doing so which hold together their different epistemic communities. To accomplish all of this, they must have and see some *mutual* goal or objective. Trade and commerce was one of the first and is still the most pervasive of these superordinate schemes. Men of different nations and of different ideologies have long ago made "one world" of disparate nations, and this will continue to be the leading edge (not the electronic media of McLuhan's "global village"). Another example is science. Less and less is the language used or the ideology involved a barrier to intercommunication among scientists of even greatly diverse epistemic backgrounds (if they are in the same or a closely related discipline!). What happens in all of these instances is that a new, emergent epistemic community is formed, one which does not require the members to renounce membership in their "back home" epistemic communities. The problem at the political level is the same: How can we agree upon a scheme which permits us to take what is said by the spokesmen of non-"friendly" nations at face value? The United Nations is but a partial answer: a man either embodies the validity of his messages for another man or he does not. The threat of retaliation does not engender trust at the communicative level.

3. There is a third origin of communication effectiveness.

119

If one controls the consequences of understanding or not, believing or not, doing or not, then one has the power to communicate effectively. All forms of socialization, from the most primitive to the most sophisticated, depend upon this kind of power. It may be exercised positively or negatively, of course. A mother may say (or imply), "I will love you if you do this or that." Or, she may say, "I will spank you if you don't." And thus the child is in-formed, not only at the message level, but at the meta-message level: It pays to do this, not to do that, and so on *ad infinitum*. It should hardly be a curiosity to us that, in traditional societies (and of course this is breaking up everywhere), it is the student who must understand the teacher, the private who must understand the sergeant, the employee who must understand the boss, and not vice versa. There is no magic in the "message." But this is how a communication model which does not include these central factors misleads us.

The traditional study of communication effectiveness leads down a blind alley. Not only must the implications of such facts as those above be accommodated by any theory of (human) communication systems but, because they are *human* communication systems, we will need to concern ourselves with criteria which are external to the process itself.

It is one thing to assess the effectiveness of a mother's communication with her child. But if her effective communication with the child impoverishes him psychologically, that is quite another thing. For the welfare of human beings *qua* human beings (in the Maslowian sense), the long-range issue may not be that of teaching people how to say, "Will you marry me and reproduce," but how to say "No."

There are local system criteria, like viability, optimum openness, etc., but every communication system is a part of the larger whole (e.g., the society), and therefore affects it in some way. So there are higher orders of criteria which must also be considered. "Romantic" love may be fun, but it may be destructive to the persons involved, or demoralizing to the society in which they and others must live. Hitler's communicative tactics were effective. But were they efficacious for the German society as a whole?

SUMMARY AND CONCLUSIONS

There are a great many other issues to be considered—such as the differences between "evolved" and "contrived" human communication systems, the fallacy of analogizing physical systems concepts like "entropy" to human communication systems, the implications of symbiotic vs. synergistic human ordering, the discontinuity between the "stuff" of human communication systems and the "stuff" of physical and biological systems, and so on. But what I have attempted to do here in this limited space is to sketch out a framework for a theory of human (inter)communication systems, and to suggest, where appropriate, the convergences and divergences of human communication theory with and from general systems theory.

I can only conclude that each has much to gain from the other, as they evolve and move toward their respective maturities.

7

Bertalanffian Principles as a Basis for Humanistic Psychiatry

WILLIAM GRAY

I am fortunate to have been a disciple of Von Bertalanffy since 1954, and I count myself privileged to have known him personally since 1966. He has profoundly affected my comprehension and practice of psychiatry, and in particular has made it possible for me to pursue an old interest of mine with renewed vigor, namely, the way in which a humanistic psychiatry could be developed scientifically. I believe that there are special features in Von Bertalanffy's presentation of general systems theory that encompass humanistic concerns as an integral part of their structure.[1] I believe that this particular contribution of Von Bertalanffy, as the father of humanistic general systems theory, has been neglected, and since I believe that it is perhaps the most important of his many great contributions, I have chosen this subject for my paper.

One can say with justice that to have laid a firm basis for a humanistic general systems theory is to have laid the foundation for scientific humanism, or humanistic science, since general systems theory can be considered as science in its most modern form. This is a magnificent contribution for a single man to have made.

A humanistic psychiatry and underlying psychology is, of course, part of such a development, and it is to this subject that I wish to turn my attention. First of all, of course, the notion of general systems theory itself has tremendous potentials for psychiatry. The idea that a human being is himself an organized system offers a totally new and fresh approach to thinking about such matters as mental health and illness. These must now be thought of in terms of effectiveness of system operation in terms of selected goals and values, overcoming the fragmented thinking of the past regarding such matters. The impossible task of defining mental health, still plaguing psychiatry, receives new impetus from a systems view. Mental

125

health should now become approximately definable in terms of richness of system structure and adequacy of function, in terms of desirable goals and values. A corresponding solution to the problem of epistemology has been offered by Walter Buckley[2] and Ervin Laszlo.[3]

The failure of the search for single causes for psychopathology becomes recognizable as part of the failure of traditional science, with its linear-reductionistic type of thinking, to solve problems where the degree of organization is high. Incidentally, one of the answers to the problem of why general systems theory appears to be in the forefront of modern science is the fact that we now live in a globally organized world with a high degree of interactivity, so that fragmented problems and solutions are no longer relevant. The notion that psychopathology is a system disfunction is of tremendous value in understanding and correcting such disorders. In this regard the climate of opinion in psychiatry has changed from one of severe pessimism to one of optimism, with many illnesses previously felt untreatable now responding to modern approaches based essentially on the systems view.

An equally valuable contribution of general systems theory to psychiatry rests on the insight that the human being is an organized system, suspended in multiple systems, large and small, of physical, social, economic, and cultural type, and that his mental health depends upon the effectiveness of the system operations that govern his relationship with the larger systems in which he exists. Thus the recent rise of the very effective methodologies of family and group therapy, which have brought many more illnesses into the realm of treatability, is another extension of the general systems idea into the field of psychiatry.

One would think that this was a sufficient contribution for a single man to have made to a field that is not his specialty, although such terminology is, in a way, not applicable to a generalist of the breadth of Von Bertalanffy. But there is more, and this is the foundation which Von Bertalanffy has laid for a humanistic psychiatry, in addition to a system-oriented psychiatry, and it is to this latter area that I wish to devote my main attention.

It is not simply a matter that Von Bertalanffy is a humanist, for many of the great workers in general systems theory are, but it is a tribute, rather, to his particular genius that in his outline of general systems theory certain features appear which introduce humanism as an integral part of general systems principles. Of course there will be many future additions and further developments, but I believe it is worth our time to pay attention to the built-in features of humanism which constitute the somewhat different structure of general systems theory after Von Bertalanffy, as compared to others.

I have, in previous papers, referred to the five Bertalanffian principles, and I will continue this procedure here. I will also try to outline their particular significance for a modern, humanistic psychiatry.

The five Bertalanffian principles, to my mind, include his insistence on an organismic or antireductionistic approach; his insistence that the psychophysical apparatus is characterized by primary activity that is antirobotic; his demand that an adequate general systems theory must concentrate on those characteristics peculiar to the human species, such as symbolism, which establishes an antizoomorphic position; his inclusion of anamorphosis and organizational laws at all levels as an essential component of advanced general systems theory, an antivitalist position; and finally, but in no way least, his demand that values, ethics, and morals are a necessary part of a new image of man and must be included in the development of advanced general systems theory, establishing an antimechanistic orientation.

The organismic position arises out of the easily confirmed observation that life occurs in discrete entities or organisms. To deny this violates biological reality. The implication of this principle for psychiatry is that it insists on a deep respect for the individual as an inviolable entity. Its importance is clear from the fact that many systems theorists have as their goal the optimization of overall system function, with little or no regard for the individual. It would seem that such approaches are, in the long run, doomed because they ignore biological reality, although it is also true that they may appear to succeed in the short range.

From my experience as a psychiatrist I am deeply aware of the psychological implications that result from such short-sighted and basically unhumanistic approaches in the form of deep anger and resentment, depression, obsessive defenses, crime, juvenile delinquency, school dropouts, the loss of a sense of meaning, and at times frank schizophrenia.

It would seem to me that awareness of primary systems orientation and utter distaste for it has spread very rapidly, so that humanistically neutral systems theory and practice would probably not be successful even in the short range. The beauty of Von Bertalanffy's organismic principle is that it in no way denies the system nature of the world, but simply places a priority on the biologically based organismic nature of life.

Although organizations are organismic in character, in a biological sense they are not alive. Of course the organismic insistence on the priority of the individual must not exclude the necessary concern for the healthiness of the ecology in which man lives. To do so would repeat the error of the Judeo-Christian view of the world as designed totally for man. Such views can be considered as humanistic, without the benefit of science. The poverty of an isolated humanistic approach is recorded in the failure of such obviously humanistic endeavors as religion and philosophy to solve the problems of poverty and illness. The future of the world seems clearly to belong to a combined scientific and humanistic approach, and I believe that general systems theory does offer this possibility.

The second Bertalanffian principle from which humanism stems is Von Bertalanffy's insistence that the essential characteristic of the psychophysical organism in all of life is primary activity. This is opposed to the previously held model of the organism as reactive, i.e., acting only after a stimulus has been applied. Such is the stimulus-response model in biology, and in psychology it exists as the pernicious behavioral school which Von Bertalanffy has justly labeled "robot psychology." The notion of the organism as primarily reactive depends upon the postulation of a homeostatic type of equilibrium in which rest, or inactivity, is the goal. In contrast, Von Bertalanffy proposes that the organism is so constructed as to depend upon a strain type of equilibrium which he has labeled the "steady state" and which, of course, makes such a system a primarily active one.

In Von Bertalanffy's model reactivity is a secondary phenomenon which allows for recognition and understanding of external reality and, to a degree, adaptation to that reality. Although adaptation is perhaps an outdated word, it is relevant mainly to the outmoded notion of the organism as a passive recipient of stimuli and information. I cannot improve upon Von Bertalanffy's brilliant description of this matter, and will quote two short passages from his works, as found in his book *General System Theory*:

If life, after disturbance from the outside, had simply returned to the so-called homeostatic equilibrium, it would never have progressed beyond the amoeba which, after all, is the best adapted creature in the world—it has survived billions of years from the primeval ocean to the present day. Michelangelo, implementing the precepts of psychology, should have followed his father's request and gone in the wool trade, thus sparing himself lifelong anguish although leaving the Sistine Chapel unadorned. . . .

Life is not comfortable settling down in pre-ordained grooves of being; at its best it is *élan vital* inexorably driven towards higher forms of existence.[4]

Here it is apparent that the second Bertalanffian principle merges with the fourth, and the steady state type of equilibrium that Von Bertalanffy proposes must be considered as a step process, with the trend toward the establishment of higher levels of development. I quote again:

In contrast to the model of the reactive organism expressed by the stimulus-response scheme—behavior as gratification of needs, relaxation of tensions, reestablishment of homeostatic equilibrium, its utilitarian and environmentalistic interpretations, etc.—we come rather to consider the psychophysical organism as a primarily active system. I think human activities cannot be considered otherwise. I, for one, am unable to see how, for example, creative and cultural activities of all sorts can be regarded as "response to stimuli," "gratification of biological needs," "reestablishment of homeostasis," or the like. It does not look particularly "homeostatic" when a businessman follows his restless activities in spite of the ulcers he is developing; or when mankind goes on inventing super-bombs in order to satisfy "biological needs."

The concept applies not only to behavioral, but also to cognitive aspects. It will be correct to say that it is the general trend in modern psychology and psychiatry, supported by biological insight, to recognize the active part in the cognitive process. Man is not a passive receiver of stimuli coming from an external world, but in a very concrete sense *creates* his universe.[5]

It is of course obvious that the second Bertalanffian principle in isolation would not serve humanistic goals, but since a necessary part of the definition of humanism is a recognition and regard for man's biological characteristics, this second principle does form an essential part of the humanistic approach. It forms the cornerstone of humanistic psychiatry. The practice of psychiatry offers abundant evidence of man's favorable response to an approach that centers its attention on his potential for independence, creativeness, spontaneity, and playfulness. It also offers the practicing psychiatrist the opportunity to observe again and again man's unfortunate tendency to behave as a stimulus-response machine, although with serious psychiatric complications when he is so treated. Psychoanalysis—and it must be recognized that Freud was a great systems thinker—has stressed, in setting up the psychoanalytic situation, the crucial importance of spontaneity rather than stimulated response. Obviously the second Bertalanffian principle is anti-robotic in character.

The third Bertalanffian principle insists that the most advanced types of general systems theory will include those characteristics that are the specific possession of man as a species. He has chosen symbolism as the primary representative of this class of characteristics, but one could add others, such as man's awareness of himself, the fact that his perspective extends beyond the range of his life, and the fact that he alone of living creatures carries an advanced awareness of the inevitability of his own demise. Von Bertalanffy feels strongly that the previous sin of anthropomorphism has been replaced in modern times by zoomorphism, represented by the trend toward believing that pigeons, rats, or other animals can be used as total models for human behavior.

As practitioners of psychotherapy, we are deeply aware that love, anger, and depression are specifically human characteristics which are supported by extensive symbolic superstructures. Even "territoriality," which clearly seems to be a factor in human affairs, must be dealt with as a richly complicated matter in the human or it is of no help to the psychiatrist. The conflicts in human beings that are relevant to psychiatry are the ones that belong to the symbolic universe although superimposed upon a biological base.

Included in this third Bertalanffian principle is consideration for what is known as "the human condition" in the work of the psychiatrist. Such psychiatrists as Seymour Halleck, whose recent book, *The Politics of Therapy,* I am happy to say, recommends an essentially humanistic general systems theory approach for psychiatrists, advises, for example, that the beginning of a definition of the human condition, for use in such an approach, must consider the human's need for intimacy, influence, freedom, openness, action, search for meaning, privacy, hope, stability, and nonviolence.[6] Another psychiatrist, L. Jolyon West, has attempted to define what he feels are specifically optimum human conditions, in what he refers to as "biosocial humanism."[7] He would include such characteristics as empathy, foresight, compassion, wisdom, judgment, insight, and love. The third Bertalanffian principle tends to merge with the fifth.

The fourth Bertalanffian principle states that humanistic general systems theory must insist upon the openness of systems and their capability for anamorphosis, that is, for evolutionary development. In addition it postulates that organizational laws at all levels exist in the natural universe and must be sought for. The fact that simpler laws do not explain the fullness of activity at higher levels is very well put by Aldous Huxley in a letter to his brother Julian, written after reading Von Bertalanffy's *Problems of Life*:

How paradoxical it is that when life develops organizations complex enough to be capable of thought, the emergent mind should revert, in its always oversimplified abstractions and generalizations, to patterns of symbols comparable in their subtlety and complexity only to organizations in the inorganic world and not to those in the living universe. . . . Hence, of course, the mess in which we find ourselves.[8]

For the psychiatrist it is obvious that organizational laws for groups and families are different from those for individuals, and that the organizational laws present flexibility and variety. The basic ideas of anamorphosis and open systems are vital to psychiatry and have rejuvenated its approach. It is surprising to note the degree to which a closed-system model has dominated psychiatric thought, in the form of assumptions that what was inside a person was fixed at birth. The notion of anamorphosis means to psychiatrists that fundamentally growth and development are a continuing process as long as life lasts, except in the presence of severe psychiatric, social, physical, and cultural pathology. The work of Rizzo in education is important in this regard.[9]

The fifth Bertalanffian principle deals with the need for the inclusion of humanistic values in systems plans, and leads to what Von Bertalanffy refers to as a new image of man in his classic book *Robots, Men and Minds*. His view is that values, ethics, and morals, although abstract, function in a very real way in the operation of concrete systems of any degree of complexity. Such values usually are subtle and not clearly expressed, and it clearly appears to be time to apply them in more explicit fashion.

The very concept of system, with its iterative looping, makes it possible to approach this most difficult of human problems with a new optimism, since values no longer have to be selected for long periods of time and will be subject to review, as are other aspects of system function. This year we will attempt to carry out such a review program at the District Court in Malden, Massachusetts.[10]

The fifth Bertalanffian principle merges with the third and adds the dimension of values that are better for humans or worse for humans, and places values on an equal footing with other system components. Since the psychiatrist has always worked with values, more systematic and conscious attention to this area should be of great help. The notion of moral neutrality is rapidly becoming recognized as a suicidal and homicidal act. Rizzo has commented that one of the signs of greatness in Von Bertalanffy is that he has never been deluded by moral neutrality.[11] Halleck pursues a similar theme in exposing the myth of political neutrality.[12] Admittedly neutrality had its value at a time when the notion of frequent review as an integral part of system function, with appropriate feedback, was not part of man's concept of the nature of the universe.

These, then, are the five Bertalanffian principles. They will, of course, be added to and further developed. For example, a humanistic systems theory should provide for the system to have available its past history, in terms of stored information. But even as they stand, the five Bertalanffian principles offer a solid basis from which a humanistic general systems type of psychiatry can evolve.[13]

8

*The Significance of
Von Bertalanffy
for Psychology*

NICHOLAS D. RIZZO

BASIC ideas with even a modicum of truth have a way of surviving. Sensational or extravagant concepts may generate more heat for a short time, but their survival depends not on the immediate enthusiasm stimulated but on the grains of truth they contain. Organismic psychology has never been associated with sensationalism. It has been a derivative of broad organismic thinking, part philosophy, part cosmology, part biology, and is based on a series of hypotheses or axioms which, though at times difficult to comprehend, are not exaggerated and meet the severest test of common sense.

The word "organismic" in organismic psychology has always been synonymous with the word "gestalt" in gestalt psychology. The German word "gestalt" means the way a thing, or experience,[1] or phenomenon, has been "put" or "put together," referring of course to pattern, shape, configuration, or form. Apparently, there is no single English equivalent to the German word "gestalt." As a school of psychology, or a system of psychology, the gestalt principle was conceived in Austria in the late nineteenth century. Its philosophical roots are much older. The so-called gestalt principle states unequivocally that the whole is greater than the sum of its parts; or, to put it differently, an analysis of parts or components cannot provide a total understanding of the functional whole.* The whole is used as the point of departure in explaining the function of parts and their interrelatedness. There are certain "whole" properties which cannot be deduced from the study of the isolated elements involved. Of course there are wholes which are mere aggregates of isolated elements, but the introduction of another concept later in this presentation will clarify the paradox.

The emergence of gestalt psychology in Austria and in

* See Laszlo, and Rosen, above [Ed.].

Germany was in part a reaction against the obviously mechanistic, atomistic, and fragmented approach characteristic of the teachings and psychological doctrines attributed to the influential teacher, Wilhelm Wundt. But organismic thinking, or global concepts, are to be found in ideas attributed to some of the great teachers of antiquity. Here, of course, we are dealing with man's struggles to comprehend the cosmos and his relationship to that cosmos. That the universe is organized in orderly fashion was the basis of the early thinking expressed by Empedocles, Heraclitus, St. Augustine, and St. Thomas Aquinas. Von Bertalanffy himself has written several essays on Nicholas of Cusa affirming the broad scope of Cusa's vision. The beginnings of gestalt psychology should be credited to the three German psychologists, Max Wertheimer, Wolfgang Köhler, and Kurt Koffka. Their early writings contain nearly all the basic principles of gestalt or organismic psychology. Metzger in Germany and Wheeler in the United States both were successful adherents and contributing pioneers to organismic psychology. Other leaders of note in the development and dissemination of the insights of gestalt psychology include Kurt Lewin, Harry Helson, W. D. Ellis, M. Sherif, and S. E. Asch. The most recent psychologists of recognized stature and influence whose theoretical bias is organismic are Piaget and Bruner. The late Heinz Werner and Abraham Maslow were of similar theoretical persuasion, especially in the humanistic core which was characteristic of their works.

Von Bertalanffy arrived at his philosophic conclusions and psychological insights, both of them organismic in character, after many years of dedicated study and research in theoretical biology and related disciplines, including physiology, morphology, mathematics, growth, and metabolism. It was due to his genius that he saw the principles of organismic psychology as special instances of the theory of a general, open, and living system. The system concept made it possible to include the heretofore so-called "nonrespectable sciences" such as psychology, the social and the behavioral sciences, under the broad theoretical framework of general systems theory. At last, the science of psychology has been accorded its rightful place beside the other sciences which deal with man.

The Significant Issues of Modern Psychology

What should the study of psychology include? What groups of problems belong rightfully to psychology? According to the *Random House Dictionary of the English Language,* psychology is defined as "1. The science of the mind or mental states and processes; the science of human nature." Other definitions include "the science of human and animal behavior and the sum of the mental states and processes of a person or of a number of persons, especially as determining action, i.e., the psychology of a soldier at the battle front." It is, perhaps, most useful to take up the problems in logical groups as they elucidate the science of mind and contribute to our understanding of the nature of man. The study of psychology forces one to decide which types of logic are acceptable. Concepts such as monism, holism, and dualism must be explained. The mind-body problem must be faced squarely. Is mind arrived at mechanistically and atomistically, through a process of additive hierarchies? Is the ultimate definition of the range, scope, and power of mind measurable in terms of the number of associational connections represented? Teleology can no longer be ignored. The *deus ex machina* of vitalistic thinking is no longer accepted.

The question of part-whole relationships has taken on added significance during the past thirty years because of the awesome potential which can be built into certain computers. It would appear that not only horizontal hierarchies, but also vertically expanding intelligence can be built into certain machines. In psychology the whole is always primary and is a unity, or integration, at whatever level it is examined. Growth and development are expansions of earlier unities and organization is assumed. The study of the processes of individuation, emergence, and differentiation form the very basis of developmental

139

psychology. The study of forms, structures, and patterns of experience is crucial because in organismic psychology it is function which determines structure in conceptual form. The manner of analysis is vital and must be nondestructive and therefore functional and multivariable in scope.

Great and lasting arguments have occurred during the past half-century over the assessment of hereditary factors in human psychology. The emergence, finally, of laboratory methods to substantiate the science of genetics has dispelled many myths. Psychology must concern itself with a study of the substrata of mental life. It has long since rejected the deterministic views which listed great numbers of instincts as fixed and specific. The same applies to reflexes and conditioned responses. The concept of straight-line causality or single-factor cause-effect relationship is no longer tenable. Field properties are among the newer concepts which account for the behavior patterns the older psychologies ascribed to the fixity inherent in instincts, reflexes, and conditioning.

Psychology, sooner or later, must face the problems associated with learning processes and their countless vicissitudes. Principles of learning, growth and learning, the differentiation process, memory function and learning, levels of insight, and the upward expansion of unity are some of the basic issues encountered in psychology.

If one pursues organismic thinking, it is not possible to avoid certain questions which, strictly speaking, are not psychological but are more or less philosophical and cosmological. If one believes in a random universe without order, then one need not be concerned about the discovery of unifying principles or isomorphies. There is, however, a high degree of order in the universe, especially in the biological world, and Von Bertalanffy has pointed out to us, through nearly a half-century of productive scholarship, the nature of that order. Organismic psychology does not exclude teleology. It attempts to discover laws of human dynamics that are not reductionistic or confined to zoomorphic issues.

Organismic psychology specifies its own methodology. Thinking may proceed by analogy if it is relativistic and perspectivistic. Analytical processes must be functional rather than structural and reductionistic. Von Bertalanffy rescued biology from its atomistic shackles at a time when only physics was considered truly scientific and in a class by itself. To psychology he has brought a series of principles which have elevated it beyond providing mere strategies for manipulation without a scientific base for further study.

Toward a Blueprint for a Newer Psychology of Man

Although Von Bertalanffy has written a highly provocative book on the current status of psychology, it is inaccurate to claim for him priority in pioneering the organismic viewpoint in psychology. What he has done is to bring to psychology nearly a half century of scholarly research in biology and related fields. His greatest contribution has been the discovery of the concept of general, open, and living systems. The human personality, with its almost infinite capacity for upward growth and expansion and its unique ability to master symbols, has served as the prototype of a general, living, open system.[2]

As indicated above, the traditional psychology of such leaders as Wundt, Thorndike, Titchener, and Watson was mechanistic, atomistic, and even mystical in conception. These psychologies did not begin with unity, but only hoped to build unities, or wholes, through principles of association and addition. The basic element was the stimulus-response bond. Teleology was denied. Although William James attempted to construct a holistic psychology, there was at the time no concurrent organismic biology for him to draw upon. Instead, his psychology relied on classical vitalism with a ghost-like entelechy making the whole, or the unit system, operate.

The natural result of late nineteenth-century European psychology, transplanted to America, was behaviorism and all its derived forms. The psychology of behaviorism, or of neo-behaviorism, is probably the most widely known in the civilized world. It is a psychology derived from the principle of conditioned response, no matter what its adherents call it. It uses models which apply equally to pigeons, rats, apes and humans. Behaviorism regards man as a reactive unit, an organism to be stimulated from the outside.[3] The clear intent of behavioristic psychology is to manipulate, be it to coax a person to buy a certain product, to feel a certain way, to accept killing, to return to hazardous duty, or to accept certain beliefs. Behaviorism denies man his soul, excepting as it can be programmed from the outside. It denies that man's most valuable trait is his capacity for spontaneous activity, that he can be guided from within regardless of the pressures exerted from without.

Von Bertalanffy's recent writings oppose certain principles of psychoanalysis, but only where psychoanalysis reduces all mental activity to sexual motivations. Insofar as psychoanalytical principles are employed to map out unconscious motivations their value remains fairly solid. Von Bertalanffy, however, was neither the first nor likely the last to point out that (1) the extravagant generalizations of psychoanalytic psychology are based on too few scientific case studies, (2) as a therapeutic modality it is pretty thin, applying to less than 5 percent of unselected cases, and (3) as a theory it is too restrictive.

The major impact exerted by Von Bertalanffy on modern psychology and on the emergence of a new image of man resulted from the forcefulness of his organismic viewpoint in biology. Similar theoretical formulations have been published in the fields of physiology, theoretical biology, growth and metabolism, economics, aesthetics, management science, sociology, developmental psychology, and the distribution of medical care. More recently, city planning and ecological problems also appear amenable to the general systems approach. But a word of warning. Although cybernetics and systems engineering have many principles in common with the theory of general systems, they have a different history. They represent a technical response to the overwhelming needs of post-World War II technology. Essentially, they are not humanistically oriented.

A modern psychology must be constructed with principles consistent with organismic biology. It must insist that mind and body are inseparable. Moreover, the whole explains the simple, or the constituent element. Wholes are always primary. A smoothly running automobile is not a whole, but merely an aggregate of assembled parts. There should be no argument here, regardless of how complex a machine may be. The human organism at birth, and even before, is an organized unity, undifferentiated to be sure, with little capacity to initiate activity, but having that capacity nevertheless. Parts or elements emerge as a result of structuring, individuation, and differentiation. The action of parts is determined by the action of the whole, according to organismic laws.[4]

The psychology of learning poses the greatest problems to the student of general systems theory and its derived organismic principles. Von Bertalanffy and many others have written frequent and serious criticisms of the mechanical models employed by Skinner and his antecedents of comparable ideological leanings. It must be recognized that teaching machines work. Additive hierarchies exist. Mechanistic principles produce learning of a sort. Operant conditioning, negative as well as positive reinforcement, are facts of psychological life. Many learners do not progress beyond the goals implicit in the mechanistic approach to teaching and learning, while some learners do acquire and develop the capacity to think independently. In many instances, an upward growth occurs with such a piecemeal approach in pedagogy, especially if the materials for study are well chosen and one deals with motivated learners. There is a great deal more for young learners to master today than was the case fifty years ago; hence the "new math" was a response (and a highly successful one) to the demand for newer conceptual models for instruction in mathematics. The knowledge explosion actually makes it imperative that newer conceptual models be developed.

Organismic laws and principles in the writings of Wertheimer, Koffka, Köhler, Metzger, Wheeler, Perkins, and Von Bertalanffy emphasize the organization and primacy of wholes. Dynamic relationships are paramount. General systems theory, more inclusive than the gestalt principle, is derived from organismic biology and theoretical biology. As a theory it includes principles which apply to the widest possible range of systems. Von Bertalanffy's genius fosters easier communication among various disciplines due to the universal applicability of his precepts.[5] Psychology is, in its broadest aspects, a study of the intricacies of the human personality not as a reactive unit, but as an organism with the capacity for spontaneous activity. Indeed, general systems theory is by far and away the most generally discussed and the broadest concept in personality theory, psychiatry, and sociology. The influence of Von Bertalanffy on American psychiatry is evident in the work of Menninger, Arieti, Rome, Grinker, Marmor, Gray, Duhl, and the writer.

In summary, Von Bertalanffy appears to have brought together the scholarship of many prior centuries in his creation of theoretical biology. With a view to man as the ultimate example of evolutionary development on earth, he has built a theory of open, general, and living systems. The theory of general systems makes possible communication across disciplinary boundaries so that knowledge need no longer be compartmentalized and fragmented. Human psychology can abandon its zoomorphic, robotomorphic, atomistic models which, in effect, have degraded, mechanized, and bestialized man. A great deal of recent work has affirmed the richness of general systems theory as a conceptual framework for the study of many clinical and social problems and of the institutions which created them.

9

Noetic Planning:
The Need to Know,
But What?

LIONEL J. LIVESEY, JR.

MY field is education. Specifically, I am a bureaucrat in planning in a state-supported university.

A few years ago, after suitable immersion in one of the better known "systems approaches" to planning and budgeting, I began to identify some serious contradictions between theory and application. What was advertised as our open search for goals turned out to be a search for techniques for maintaining the machine, which our colleague Kenneth Boulding has criticized as the concept of ". . . society as having a single well-defined end which is to be pursued with efficiency."[1] Again, what was advertised as seeing the system whole emerged at the end as just one more application of reductionism to a social institution—that is, we saw no forests because of our usual preoccupation with the trees.

Determined to find out what went wrong between the promise of the systems concept and its application, I began to delve into general systems theory.

In the first part of my presentation, I shall describe my "Pilgrim's Progress" through the land of general systems theory. In the second part, I'll explain why this is more than an intellectual exercise. As a practicing planner in the field of education, I'm obliged to look for answers to the question: *What do we do on Monday morning?*

Miller, Galanter, and Pribram, in their book *Plans and the Structure of Behavior,* restate this question in another way:

. . . It is so obvious that knowing is for the sake of doing and that doing is rooted in valuing—but how? . . . Does a plan supply the pattern for that essential connection of knowledge, evaluation, and action?[2]

I think that a planning *process* could provide for the connection. I'll suggest ways in which general systems theory might assume the role of justice of the peace in providing marriages of thought and action.

147

I

The three problems that I have with general systems theory are (1) my feeling that it is not useful [and perhaps arrogant] to construct taxonomies of physical, biological, or social systems which put man too much at the center of the universe thus perpetuating the impression that evolution exists solely for his sake; (2) my confusion in understanding whether general systems theory is philosophy or science or some other discipline, indeed, why we need to confine it as a discipline at all; and (3) my curiosity concerning the relationship between order and disorder (that is, deviation-reducing and deviation-generating properties) in systems, which is to ask how we may account for the energies of creativity, leadership and imagination.

Again, let me say that these are action-oriented questions. For example, in one scenario please picture me as teaching general systems theory to lay persons as a prerequisite to planning. I mean some purposive activity such as the redesign of a school system, the design of a model city, or development of some larger region of the ecosphere. For such purposes, I would need to place man in the right perspective. And I need to know what, if anything, remains outside my discipline. Most of all, I would need some arguments to counter the inevitable yearnings for homeostatic maintenance.

Organization of General Systems Theory

In urging that general systems theory develop a form and structure of its own, Kenneth Boulding has advanced two possible approaches. One of these is to arrange theoretical systems and constructs in a *hierarchy* of complexity, roughly corresponding to the complexity of the components of the various empirical fields. The second approach to structure is to build up general theoretical models of certain general *phenomena* which are found in many different disciplines, that is, a field theory of the dynamics of action and interaction.

Boulding prefers the first approach. His "skeleton" of "levels" of theoretical discourse consists of eight levels of complexity. These are (1) the level of *clockworks,* which includes both static and dynamic physical systems that exhibit a tendency toward equilibrium; (2) the *thermostat,* or homeostasis model, that differs from the first in that it transmits and interprets information by means of cybernetic mechanisms; (3) the level of the *cell,* the first sign of life that has the properties of self-maintenance and self-reproduction; (4) the level of the *plant,* which begins to differentiate between the genotype and phenotype, and exhibits the phenomenon of equifinal or "blueprinted" growth; (5) the level of the *animal,* characterized by increased mobility, teleological behavior, and self-awareness; (6) the *human* level in which individual man or woman is considered as a system, with the distinguishing capacities of language and symbolism and the self-reflexive quality; (7) the level of *social organization;* and, finally (8) a "final turret" for *transcendental systems.*[3]

I have seen other equally elegant models of hierarchies. In particular, I remember one taxonomy of complex adaptive systems which created three models: one for the geosphere, another for the biosphere, and a third for the noosphere, after Teilhard de Chardin. In these models evolution spirals upwards from plateau to plateau from the lowest level to the highest levels of complexity.

If I must accept the basic premise of disciplining the interdisciplinary, that is, finding form and structure for general systems theory, then I much prefer the notion of encouraging construction of a "field theory" of the phenomena of dynamic action and interaction that may be found in the empirical world.

The two approaches are complementary, of course, but my preference for the latter is an instinctual dislike for hierarchic models that put man either at the center of evolutionary activity or at the top of the evolutionary pyramid, as the case may be.

In the case of Boulding's hierarchy, I get a glimpse of movement between some of his levels over evolutionary time, but I get no sense of dynamic interaction among all his systems *at a given point in time.* I have the same trouble with the charting of the empirical world by modeling the geosphere, the biosphere, and noosphere into self-contained taxonomies of upward-spiraling complexity. Again, how would the spheres interact in a cross section of life and nonlife, now, or at a given moment *in the future* which is the stuff of planning for purposive action?

I am quite willing to admit that I'm conditioned by my upbringing on an isolated ranch in the Colorado mountains. In such a setting one literally does not think of man as being isolated from his environment, or as superior in terms of "conquering" all other so-called "lower order" systems. For survival's sake, one *comes to terms* with plants, fish, birds, animals, and the formidable forces of wind, rain, fire, snow, solar heat, and the other presences of geology and climate that make life possible. Instead one has those feelings of kinship and respect that color the culture of the Sioux, the Navajo, and other American Indian nations. In such cultures, the creeping people, the standing people, the flying people, the swimming people, and the forces of air, water, soil, and fire are incorporated into councils of government. The Indians knew that if man tried to insulate himself from nature, the other systems of nature would be capable of making nonnegotiable demands. We are now somewhat tardily discovering this truth.

Am I guilty of anthropomorphism? I deny the charge, should it be lodged against me. The last thing I should ever want to do is to attribute human characteristics to the other orders of living and nonliving things. On the contrary, I would accept the possibility that "they" may have ways of "knowing" the world that are different from ours.

Am I guilty of denying man his rightful place at the pinnacle of evolution? Perhaps I am, but I do it with an ear to the comments of some of our young people. They already understand that man, through his fantastic development of tools

150

that extend his senses and do work, has gained such an advantage over his environment that he has nearly reached a point where he can decide what he and his world will become. But my young friends also see this as having been an exercise in arrogance and exploitation. They search for the restraint and sense of responsibility which they regard as a prerequisite to the survival of all systems, including man.

All I am talking about, really, is an attitude. If I were ever asked to teach general systems theory to today's high school or college students, I think I'd steer clear of hierarchies that, consciously or unconsciously, create the impression that evolution was invented for the sake of mankind, even though it *is* recognized that further evolution could conceivably be controlled by man. By teaching a field theory of dynamic interaction between *co-existing systems,* I might circumvent the present distaste of youth for the *hubris* of the technological man. Together, we might grasp the full significance of Theodosius Dobzansky's call for a new attitude, namely the truth that ". . . by changing what he knows about the world man changes the world he knows; and by changing the world in which he lives man changes himself."[4]

Science or Philosophy?

Let me now present my second problem, what I defined before as an attempt to understand whether general systems theory is a philosophy, a science of sciences, or both, and what, if anything, remains outside this discipline of disciplines.

This is a confusing matter for an amateur. I can only regurgitate what I have learned and ask to have it clarified a little.

Apparently the first battleground for general systems theory consisted of breaking with the "physicalistic" approaches of the classical sciences, that is, the world view of linear causality and reductionism which (in Von Bertalanffy's words) assumes ". . . that reality is 'nothing but' a heap of physical particles, genes, reflexes, drives, or whatever the case may be . . . ," and that the whole is "nothing but" the sum of its parts. This break called for an understanding that ". . . not only the elements but their interrelations as well are required: say, the interplay of enzymes in a cell, of many mental processes conscious and unconscious, the structure and dynamics of social systems and so forth, [which] . . . requires exploration of the many systems in our observed universe in their own right and specificities . . . [while] it turns out that there are correspondences or isomorphisms in certain general aspects of systems [which are] the domain of general systems theory."[5]

Lately it seems that the break with classical science has become more sharply focused in a further rejection of the "zoomorphic" view as being ". . . the 'robot' model of man, which considers the psychobiological organism essentially as a machine responding to stimuli and governed by utilitarian factors. . . ."

In this sub-battle of the war against reductionism, it is not difficult to support his attack upon a "materialistic and commercial society" that favors ". . . a patently insufficient 'model' which 'explains' Leonardo's paintings as outflow of an infantile Oedipal shock, and language with all its 'meanings' as Pavlovian or Skinnerian conditioning. . . ."[6]

I have learned that general systems theory is a "perspective philosophy." What this means to me is that general systems theory has moved into a territory once reserved to classical philosophy—at least, has occupied a *part* of classical philosophy's formerly exclusive right to undertake "theory or investigation of the principles or laws that regulate the universe and underlie all knowledge and reality."

I can follow the line of reasoning which disposes of some of the perennial (but sterile and anemic) problems of classical philosophy such as the central one of the nature of the mind-body relationship, "I" and "it," subjectivity and objectivity. I can grasp the notion that since the mind is part of nature, we might overcome the problem of worrying about whether some structure exists in reality or is simply a projection of the human mind. I welcome the potential use of general systems theory in

construction of a process-model unifying subjective and objective experience.

This small amount of learning on my part, however, leads
to two sets of questions—one set which is relatively superficial,
and another which is more substantial.

The set of superficial questions has to do with labels. Some
systems theorists refer outright to GST as "science," and others
refer to it as a "perspective philosophy." I would imagine that
the resolution of this apparent difference in terms is merely
that general systems theory is a "science of sciences" which fits
the dictionary definition of "philosophy." So far, so good, except that we are reminded by Laszlo that ". . . Not all of science
is relevant to philosophy, and not all of philosophy is relevant
to science."[7] Are we not, then, back where we started from?
What, if anything, in *science*, is *not* embraced in general systems theory? And what is the residue in *philosophy* that is *not*
embraced in general systems theory? If general systems theory
is, indeed, a "discipline," what name should it be given to differentiate it from the things with which the founders may not
wish to be identified? On the other hand, if general systems
theory *is not* a "discipline," what difference does it make?*

Let me proceed immediately to my second set of questions
which explores the same confusion, but at a different level.

I referred earlier to the mind-body problem, and the idea
that objective and subjective experience *might* be unified as
one process, and that *maybe* a way will be found someday to
discover what goes on in a scientific sense—that is, discoveries
of a law or laws that can be verified empirically. But the fact
is that the relationship between "private" and "public" experience is presently *not* known. And the possibility also exists
that the truth is unknowable in terms of science or philosophy
or any rational terms. Professor Von Bertalanffy has said quite
recently in an article, "System, Symbol and the Image of Man"
(which I commend to everyone for its clarity, elegance and

* Here cf. Laszlo, above [Ed.].

economy of thought), that ". . . what 'reality' ultimately is behind the phenomena of both subjective and objective experience remains beyond the limitations and even the interests of science. However it may well be that we have other access to this reality in direct experience and its sublimation in art, music, and mystical knowledge."[8]

I am intrigued by the word "interests" in the phrase ". . . beyond the limitations and even the interests of science." Does this mean beyond the interests of classical science only, or all science including the current process-oriented science? Does this also mean beyond the interests of general systems theory?

Meanwhile, pending enlightenment, let me return to the little scenario in which I wondered what I would do if I were to teach general systems theory to the high school and college students who *say* they are "turned off" by science and philosophy, classical or otherwise, because of "bad vibrations" that seem meaningful to them. Until I became more sure of my ground, I think I would say to them that general systems theory is not a "discipline" in *their* understanding of the term as a "structuring and partitioning" of information. I think I would tell them that it is a kind of wisdom that illuminates all disciplines, and yet does not pretend to be complete. I would remind them that general systems theory is *one of many* useful ways of knowing, *not greater and not less than* the revelations that may come through "art, music, and mystical knowledge," and other unverifiable and nonreplicable experience. I would tell them we need all the ways of knowing we can find.

Leadership, Creativity, Imagination. We say that the system known as man is a "primary activity," rather than a "reactivity" in the cybernetic sense, or in the sense of stimulus-response operant conditioning. We say that a social institution is a network of interacting processes which is capable of evolving into ever more elaborate networks.

In all the models and flow charts I have studied I can identify the negentropic process in which new energy enters a system from outside. That is, I can see the principle by which a social institution, for example, is renewed as new "information" is introduced (alterations in the system's environment, introduction of new members to a system's population, transfusion of new purposes, and so on). But what I fail to find in my limited acquaintance with general systems theory is some sort of principle that admits the *self-renewal* of a social system by means of some internal process that creates change. The notion that a social system exercises "selectivity"—that is, *accepts* new energy with which it feels compatible, or *rejects, coopts, or modifies* disturbances which threaten its tranquillity —does not exactly provide an answer to my question.

The fact that I have not found what I am looking for in general systems theory does not mean it isn't there. It is more likely that I haven't found it yet.

Meanwhile, I do have bits and pieces that do not fit. On the discouraging side I am told that the exercise of free will— that is, an original course of action created out of the imagination of "great men" or "rugged individualists"—cannot be regarded as a quantum leap of self-renewal originating in a system; rather, such visions and creators of new visions ought to be conceived as acting like "leading parts," "triggers" or "catalyzers" for something good or bad that was bound to happen anyway.

Against my will, I fear this may be true, which means to me, as planner, that all change must be introduced from the outside in. And yet, I still hang on to some encouragement from two authors I have read.

I think Karl Deutsch admits that there may be some sort of impulse within a given system that deliberately seeks out disorder and creative chaos, even as other impulses in the system seek to bring arbitrary deviation back to a normal curve. Deutsch, who speaks in terms of cybernetic models, thinks that ". . . the highest order purpose in a feedback net . . . *might* include states offering high probabilities for the *preservation of processes of purpose seeking,* even *beyond the preservation of any particular group or species of nets* [italics mine]."[9]

155

Again, Walter Buckley seems to admit the *need for deviation* in mental and/or social processes. I like his call for assessment of ". . . the ongoing process of transitions from a given state of the system to the next in terms of the deviation-reducing and *deviation-generating* feedback loops—relating the tensionful goal-seeking, decision-making subunits via communication nets . . . [my italics]."[10]

I hope we can identify a kind of "rogue event" in systems that is itself preserved. I'll not belabor the idea of a property of systems that creates or searches out disorder, except to mention that if, as it is said, our values are forged "in the conflict of opposites," then our disorder-seeking impulses are needed to ensure a *ready supply* of opposites. What worries me, as will be apparent in a moment as I turn to a scenario to "educate for planning," is that there is a growing shortage of creative chaos in this world. The "advance men" for 1984 are enjoying much success in promoting the utilitarian utopia. We may well get there ahead of schedule.

Meanwhile, as I close this first section of my presentation, let me return once more to that hypothetical situation in which I might find myself teaching general systems theory to high school or college students. How shall I answer their questions about the place of the imagination and the role of creative leadership in the systems view of the world? It seems now that I have two choices. I can say that general systems theory encompasses all aspects of transition from a given state of system to the next. Or, I can fall back on what we have said before which is that there are other ways of creating a reality through "art, music, and mystical knowledge." This means that planning is not a science but an art informed by intuition.

II

At last I've come to the concluding, and much shorter, section of this paper which is one man's answer to the question:

Noetic Planning: The Need to Know, But What?

What do we do on Monday morning? I am sure you understand
that my work as planner in the field of education is such that,
whether or not I am ever able to comprehend the full breadth
and depth of general systems theory, and make the best use
of it in educational reform, I am still obliged to act.

The initial step in planning which is very often overlooked
is to study the *present,* which someone has described as being
"borne of the past, and pregnant with the future." The device
I favor is to construct a *collage* of the "booming, buzzing confusion" in which we find ourselves, and to examine it for any
patterns that may emerge.

Fortunately a good *collage* has been put together for me,
for my present purposes, by William Irwin Thompson in his
book *At the Edge of History.*

Thompson says:

We do not know the answers yet, but I think some of the old
questions are going to drop as we let go of the illusions that make
them necessary. At the moment we are still ignorant enough to
think that there are no problems with the commonly accepted nar-
rative of history; and that is an impediment to further knowledge
that should be removed. It is difficult to move on when our ignorance is so comfortable and our intuition is warning us that the
new knowledge will unsettle our established . . . culture. And so
it will be only those who are uncomfortable with our established . . .
culture who will get up to go and see what is outside the air-
conditioned, acoustically tiled, and fluorescently lit container our
civilization has chosen to end up in.

Some of those who are getting up to go and see are insane, some
of them are hippie members of UFO cargo cults, some of them are
failures in the status quo . . . some of them are revolutionaries who
simply like destroying things for the sake of destruction, *and some
of them are the original thinkers upon whom scientific knowledge
depends.*

Some of those who are *not* getting up to go and see are sane,
some of them are failures in the status quo who identify with their
superiors to disguise their failure, some of them are bourgeois . . .
industrialists who cannot leave because the triumph of the system
is what they struggled so long to achieve, some of them are important people who like exercising power for the sake of power,

and some of them are the unoriginal thinkers who maintain the past knowledge upon which all science depends.

We are approaching a point at which we will have to choose, and the choice will be hard, but it will be good, for the choice will tell us, in an almost instinctive act, exactly where we are.[11]

There is a pattern of social behavior that emerges from Thompson's *collage*. This pattern contains several imperatives for education and for long-range planning:

1. *We are becoming a polarized society.* For some time I have been coming to the conclusion that in this country we are living, not in a time of revolution, but in something more like a civil war between habit, routine, and enclavism on the one hand ("those who are *not* getting up to go and see") and openness, possibility, and spontaneity on the other ("those who *are* getting up to go and see").

The first imperative is participation by everyone in the improvement of his capacity to know his world and find out who he is. This is no more than an affirmation of the challenge given us by Thomas Jefferson, which is part and parcel of the meaning of democracy. He said:

I know of no safe depository of the ultimate power of society but the people themselves, and if we think of them not enlightened enough to exercise their control with a wholesome discretion, the remedy is not to take it from them but to inform their discretion.

2. *We must become a noetic society.* I am indebted to James D. Carroll for the adjective "noetic." The word derives from the Greek *noein,* to perceive, which is a form of *nous,* or mind. Carroll argues that, in a psychological (and not a Marxian) sense the state is withering away. He says that ". . . what is withering is confidence in the state, as presently constituted, as an open political order for structuring processes of persuasion, bargaining and trade-offs to make legitimate . . . decisions concerning basic social conflicts."

He faults the hierarchical order and control in *all* our institutions which have become "increasingly a service mechanism for satisfying material concerns," whereas "noetic authority" should be ". . . the willingness and capacity of individuals to function in cooperative systems . . . to be reconstructed through open political processes of inquiry and search, to direct constructively the growing tension between individual freedom and public order in a changing environment."[12]

The imperative for education and for planning is to accommodate the new awareness through lateral and collegial, rather than through hierarchical, definitions of reality.

3. *Noetic Authority must be served by Noetic Planning.* I submit that the function of planning for education is not to provide means to indoctrinate the young in the current cultural tradition and value system of society in service to a hierarchical state with *"a single well-defined end to be pursued with efficiency."*

I suggest that planning for education and education for planning are the same since it is the chief business of persons, both young and old, to improve their capacities to choose, and choose again, in "open political process of inquiry and search." Under such conditions, planning becomes curriculum. We must be learning individuals who plan. We must be planning individuals who learn.

The imperatives for noetic planning are to devise new ways to learn and choose.

4. *Systems education must accompany Noetic Planning.* The reason I am naming systems education as a component to Noetic Planning is to overcome our inability to see things whole. In my plan some training in general systems theory would precede active engagement in problem solving, in much the same way as familiarity with scientific method is now taught before entrance to the laboratory.

The imperative, the task for education, expressed so well by Paul A. Weiss is to delineate ". . . boldly and in sharp relief the solid core of our present knowledge without embellishment [to] bring back the realization of the immensity of our remaining ignorance." He says we must find ". . . the conceptual integration that renders the map of knowledge not only more complete, but more consistently coherent."[13]

5. *The nurture of creativity must accompany Noetic Planning.* As I have said before, I suspect that we are already on the way to becoming the anthill society of 1984.

Part of the reason is our simple failure to deal with technological complexity, ecological complexity, medical complexity, urban complexity, and governmental complexity. Our lack of "complexity-consciousness"—and our inability to see things whole because we do not know where to begin and have no teachers who can help us make a start—led me to urge the teaching of general systems theory.

But our problem also relates to our willingness to delegate our fate to technicians who have overstressed means and neglected ends to such a degree that all of us are forgetting how to imagine anything else except servicing the megamachine. Our cult of efficiency has starved our imagination.

The imperative, as Robert Jungk observed in a speech to architects and engineers, is ". . . to think about these millions of dried-up imaginations in millions of individuals and create conditions that will make them come back to life."[14]

In conclusion let me come to one scenario for Noetic Planning that might give a bit of specificity to what I have said.

Although I have chosen an example of my different approach to planning on a *national* scale, I could just as well refocus it on a local institution: a school district, college or university, city council, state government or other.

Imagine if you will a "school for planning" in Washington, D. C., designed to serve the Congress by exploring certain megaproblems before they reach the crisis stage. Let us assume a faculty of comprehensivists who come and go as needed, Chatauqua-style. Let us assume a student body of all ages mixed together who have already had exercises in reawakening creative thought (as will be provided by the "school") as well as enough of an introduction to general systems theory to enable them to look for the isomorphisms in the physical, biological, and the social and symbolic worlds. At this stage we are still cultivating a new attitude toward complexity.

We are now ready for the laboratory in which we shall find and explore a given megaproblem, as for example, the rapidly approaching prospect of an automated world in which we'll have no further need for human work.

At this point, a variety of learning-planning styles are possible, but I have chosen the courtroom mode.

Long before we reach the courtroom stage, however, I would have divided my students into two groups: one to develop a scenario (plan) for managing a world without human work, and a second to develop what is called "a bitter enemy" scenario (or counterplan) which would oppose the first at every point. We might then be contemplating a society in which every individual is freed to devote all his time to arts and crafts as compared to one in which the work ethic remains predominant, that is, unchanged. I do not know. Exactly what the plan and counterplan might be rests with the imagination and research performed on each.

I would then bring my teams together in a public forum closely following the courtroom mode. The plan for a world without human work would be placed on trial, to be prosecuted by the counterplanners and defended by the devisers of the plan. Each side would present witnesses (expert testimony), cross-examine witnesses, file briefs of every sort. We may have a judge and jury, too, to add to the excitement, but they are not really necessary to the ends I have in mind.

I would hope that my students would have received a holistic education beyond anything available elsewhere. I would hope that their *in-formed* creativity would have imagined a display of alternatives for choice that is not presently available in any other planning mode. I would hope that the public forum would produce a "testing of values in the conflict of opposites" that would facilitate the emergence of new

LIONEL J. LIVESEY, JR.

values to meet new conditions. Above all, I would hope that the process would demonstrate what Karl Mannheim called ". . . the real difference between dictatorial and democratic planning."[15]

In closing, let me offer a quotation from Jurgen Moltmann which accomplishes two things. It gives concreteness to the notion of *forced free time* which I have chosen to put "on trial" in my plan for planning as described above. In addition, it speaks with compassion to the need for hope and planning in the postindustrial society:

In highly industrialized lands, the signs are increasing that the category of the "future" is ending. . . . In place of what were once necessary and substantial investments in an increasingly attractive future, there is a consumer relationship which is completely even with the future. The anticipation of disappointment gives birth to all those critical questions about great plans and models of a world of tomorrow. Does planning make man the free master of his history, or does it lead him into an organized adjustment to barren and unappealing developments?

Does planning arouse a new consciousness of the future, or does it de-futurize the future? Shall we, as G. Anders put it so pregnantly "in the future no longer see the future as future?" Does planning effect the overcoming of blind fate by seeing the responsibility of men or is man, through a totally planned life, the fatalist of his own timetable? What, for example, did the programme for the shorter working week promise? An increase in leisure time, freedom and self-realization. Why did this not happen? Obviously, the nonidentity of man remains a torture and a piece of good fortune. *Horror vacui* asserts itself when men "do not know what to do." We must certainly be aware of this distinction for man if we are not to appropriate it and to produce those disappointments the consequences of which we cannot foresee.[16]

If planning is hope, then both are now on trial. As Shakespeare said, for better or for worse,

It hath been taught us from the primal state
That he which is was wished until he were.

10

The General Ecology of Knowledge in Curriculums of the Future

JERE W. CLARK

As a representative of the education community I should like to sketch in a somewhat speculative but reasoned vein, the nature of the education curriculum which general systems theory seems to make possible, and to sketch likewise the critical role this new orientation to knowledge and education might play in shaping the global village of the twenty-first century. I believe this may be an appropriate way to pay tribute to our distinguished colleague, Professor Ludwig von Bertalanffy, the "father" of general systems theory and a founder of the Society for General Systems Research (SGSR).

By way of preface I want to acknowledge my bias in this matter. I have a subjective feeling that the SGSR is located historically in a strategic position to be, if it should so choose, the innovative catalyst for shaping and getting into motion the needed education program on a worldwide basis. Although I am presenting my personal opinions regarding the needs and opportunities of the day, I want to acknowledge my indebtedness to many members of the Task Force on General Systems Education—the education committee of the SGSR—for helping to identify and clarify some of the issues involved.

I. THE GLOBAL
ENVIRONMENTAL SETTING

Breakdown of the Symbolic Universe of Values

As has already been implied, the timeliness of Von Bertalanffy's contributions to education can best be seen and evaluated against the background of the central ecological problem of our time—namely, the survival of global civilization.

If I interpret him correctly, he has defined this problem as the tendency toward "the breakdown of the symbolic universe" of values.[1] His prescription? Keeping in mind that the uniqueness of human nature "is the creation of symbolic universes in language, thought, and all other forms of behavior,"[2] the solution to this gigantic problem requires that "a new symbolic universe of values must be found or an old one reinstated if mankind is to be saved from the pit of meaninglessness, suicide, and atomic fire."[3]

The Need for a Challenge

The penetrating power of Von Bertalanffy's X-ray mind in the sociosphere is illustrated by his diagnosis of the present cultural crisis. He says that our society in America today is tending to become neurotic, not because biological survival is at stake, but because we as a society are not responding to a great enough challenge. "When life becomes intolerably dull, void, and meaningless—what can a person (or society) do but develop a neurosis?"[4]

Nature of the Challenge

Could it be that we, as a national and global society, have a twofold challenge emerging from this definition, diagnosis, and prescription of the problem? If so, the first part of the challenge could be identified by viewing general systems theory as a possible—perhaps the best?—conceptual vehicle now available for delivering "the way," or the road map, showing how to transform today's world into the desired world of tomorrow. The other part of the challenge could be identified by viewing general systems *education* as a possible—perhaps the best?—educational vehicle for delivering the general systems "message" to those who are to use it in charting the way to the future.

Toward Grass-roots Involvement

Von Bertalanffy has built into the prescription another condition which seems to multiply the size of the challenge severalfold while making it also more realistic. Here is a clue:

166

". . . Culture, that is, a framework of symbolic values, is not a mere plaything for the human animal or luxury of the intelligentsia; it is the very backbone of society. . . ."[5] In other words, in a participative, technetronic democracy where success depends on getting everyone into the act of planning for the future, there seems to be little basis for hope in the outcome unless there is a common language and common orientation to the problem. In terms of the educational dimension, this means that our delivery system, at one or more levels, must reach not only the relatively few intellectuals and general systems practitioners of our day but also the masses of citizens as well.

Toward Harnessing the Aspiration of Youth

Given this communications bridge to grass-roots involvement in designing and finding a way toward the world of tomorrow, might there be a basis for suggesting the possibility of using these two knowledge systems together (general systems theory and general systems education) to serve as a delivery system for the more promising dimensions of the so-called youth culture? We find today's youth (in spirit and/or in chronological age) crusading for the "promised land," but with neither a real map or globe of the new world, nor a road map showing how to get from one realm to the other, or even where they are now situated (relative to the promised land). Much energy, enthusiasm, determination, imagination, and work have been invested in the crucial enterprise of alerting society to the fact that our global spaceship may self-destruct if some very basic changes are not made in direction. That being the case, it would be unfortunate for the joint effort to result mainly in frustration and cynicism for its participants and an increased chance of doom for our national and global society. Could it be that our basic need today is to rewire our conceptual, natural, and educational worlds for space-age communications linked to the new societal and environmental values which are emerging?

167

II. THE GENERAL ECOLOGY
OF KNOWLEDGE

It is the urgency of getting on with a massive educational program of general systems education which leads me to suggest that we consider using the term "general ecology of knowledge" as a supplementary phrase for describing general systems theory. Very limited experience has indicated that this term might prove to be a more meaningful and attractive label to the masses of educators and the public generally than the term "general systems theory." There unfortunately seems to be a tendency on the part of many educators and the public to assume erroneously that general systems theory is static, impractical, mechanical, linear, closed, and "soulless." Having heard so much about inflexible, quantifiable, carefully delineated hardware systems, they sometimes seem to assume that general systems theory may represent an effort to view people as cogs in the wheels of some arm-chair theorist's model of the universe.[6]

In this age of environmental crisis, it is encouraging that people generally are beginning to get the message that the emerging general ecology of nature tends to be dynamic, action-oriented, organismic, relative, nonlinear, open-ended, value-conscious, holistic, and interdependent. The hope is that we might be able to convey the idea that we are striving to develop an image of nature which mirrors nature's system of systems or general systems. Then the question becomes: Might using the terminology "general ecology of knowledge" help to make this transition?

We are using the term "general ecology of knowledge" heuristically to mean the study of patterns of interrelationships among the various "species" (subsystems, sub-subsystems, etc.) or fields and subfields of knowledge with emphasis on: (a) preserving the condition of dynamic balance between the "species" and their environment; and (b) optimizing the overall, symbiotic fruits of synergistic interactions among them.

The definition of general systems theory implied in this definition of the general ecology of knowledge is, I believe, broader in scope and more ambitious in thrust than the usual definition. It reflects the aspirations of contemporary general systems theorists as well as the achievements of the theory to date. It includes the intuitive, heuristic, analogical quest for isomorphisms as well as the final delineation and rigorous specification of the isomorphisms. Rather than being limited to the exploration of only selected types of systems and systems problems, it embraces the study of *all systems of organized energy* and, hence, seems to provide a means of orienting knowledge generally. Using it as a criterion, much of the research done by some organizations, such as the Institute for General Semantics and the Center for Integrative Education, would be included, although these groups do not generally view their work as general systems research as such. In other words, any research which either succeeds or promises to succeed in systematizing, integrating, simplifying, and operationalizing knowledge generally would be included. This conception seems to have much in common with Von Bertalanffy's "new natural philosophy" in which "a new conception of the world as organization seems to emerge,"[7] although it is probably more heuristic and somewhat more comprehensive than what he envisions.

Let us now turn our attention to general systems education and see how general systems concepts can be used to redesign our education system.

III. THE EMERGING GENERAL SYSTEMS EDUCATION

Programs in General Ecology of Knowledge

As is true in designing the program of study in any field of thought, the needs of both prospective "majors" and "non-majors" in general systems studies must be met. As is usually the case, the curriculum for the "nonmajor" is designed to provide a working orientation in the field and its relationship to other fields.

Once this overall program is developed, tested, and generally used throughout a given school or college or other institution, however, the need for even one special course or seminar would more or less disappear. Each course in the major in each traditional field such as botany, or economics, or physics would be developed within the framework of the general systems orientation. Ideally, we hope to see developed what Kenneth Boulding has called "specialists with universal minds."

We rather suspect that some one basic concept or process in any one of the main academic disciplines (ranging from the hard sciences to the social sciences, arts, and humanities) can be used as the focal point around which the entire universe of knowledge can be oriented. We also suspect that in most, and possibly in all, cases this can be a *functional*—though heuristic —orientation to the entire universe such that its workings can be "explained" to a large extent in terms of that principle. Some examples of such unifying concepts and fields are: the periodic table of elements in chemistry; the concept of $E = mc^2$ in physics; entropy in biology (or any one of several other fields); and the principle of opportunity costs in economics.[8]

A very suggestive embryo of possible programs for "majors" in general systems studies (or in the general ecology of knowledge) is outlined in the appendix to this paper.

Redesigning Curriculums Generally

In view of the fact that today's educators were reared and educated in the natural and conceptual worlds which are suddenly becoming outmoded, it is not surprising to find many schools and colleges geared more to the past than to the future. Educators generally are not to be blamed for not already having a working orientation to the emerging, functionally in-

170

tegrated patterns of knowledge and education. However, pressures stemming from threats to our civilization are building up on us and on other general systems groups to do everything possible to help educators become acquainted with the value and usefulness of the *attitudes, approach, and simplest tools* of general systems theory (within the context of its common orientation to all fields of thought, nature, life, and education).

In past decades, when the coming metamorphosis of values and global perspectives was not foreseen, it is understandable that educators assumed that the future would be essentially a simple projection of past trends. In such a more or less static world of specialization, a well-filled memory and the ability to reason logically seemed to be the main outputs needed from the education system. With the shift from the local to the global perspective, from the relatively static to the dynamic world, from independence to interdependence, and from unguided growth to the quality of life, the requirements of the education system seem to have multiplied severalfold (at least during this period of transition). Today the student must be viewed as a potentially *self-propelled, self-guided navigator* moving not only through the relatively familiar, static, and simple terrain of today but also into the relatively strange, dynamic, interdependent, and complex world of tomorrow. His mind must be programmed so that he can reprogram it whenever necessary throughout his lifetime.

In the emerging general systems education programs an effort is being made to select knowledge inputs which are valuable, frequently needed, intriguing, measurable, "retrievable," versatile, and relevant to the world of tomorrow. Instead of concentrating primarily on reason and memory—functions which the computer can do better than any human—the new program concentrates on generating the optimum blend of such problem-solving skills as the ability to observe (especially to recognize patterns), communicate, value, imagine, reason, hypothesize, measure, model, select, plan, test, implement, and reevaluate.[9]

171

The urgency and scope of the need for some form or variation of the new type of educational program throughout the world, combined with the simplicity and self-directing, self-propelling nature of the program, suggest the need for a *mutually reinforcing, two-track strategy* for delivering it to the world. One track would be through the traditional educational institutions whenever and wherever possible or feasible. Because of the difficulties normally experienced by large institutions, such as our education system, in responding as a unit to the need for change, however, the main thrust initially might be made on the second track which is administered independently of regular schools and colleges. This means that the emphasis initially would be on developing materials and learning aids which any literate person of any age would use on his own, at his own speed, and for whatever number of hours per week he might choose. Professional teachers who want to get in on the ground floor of the movement, but whose administration could not provide the needed preparatory programs, could move along this track until the other track had been established in his system.

The Beginnings of a Case History

The role of general systems theory in restructuring and revitalizing educational curriculums can be illustrated to a limited degree by describing briefly the experience of a collegiate center which was set up in 1967 specifically for that purpose. This is the Center for Interdisciplinary Creativity, which originally had been set up as a Center for Economic Education, at Southern Connecticut State College.

The entire educational and research program at that Center was oriented around the following words from Ernst Mach in *The Science of Mechanics*: "The purpose of science is economy of thought." The broadened mission of the Center was to apply the generalized version of the economizing tools of the economist (such as the principles of opportunity cost and substitution) to the twofold task of restructuring knowledge and

172

curriculum patterns to meet the requirements of the space age. The basic research arm of the Center was thus commissioned to use these economizing principles to restructure and unify the social, biological, and physical sciences, in much the way an innovative businessman would restructure or modernize his plant and operations in the business world. Using essentially the same principles, the curriculum-development arm of the Center was commissioned to restructure the educational curriculum at all grade and age levels from kindergarten through postgraduate and adult education.

A dynamic, heuristic paradigm integrating the social, biological, and physical sciences (and the value system from the humanities) is being developed to serve as a guide to both pure research and curriculum research at the Center. This model is being developed by fitting a generalized extension of the cybernetic process (i.e., a "paracybernetic" submodel) *functionally* into the context of the generalized economizing process.[10]

The rationale of the "economizing" approach to unifying and simplifying knowledge, curriculum, and learning patterns might be couched in terms of three major concepts from economic history. One provides a suggestive means of unifying and simplifying the structure of knowledge in the social and natural sciences. Another provides a basis for adding the dimensions of relevance, sensitivity, coordination, and effectiveness to the curriculum. The third provides a basis for developing mental wings with which the average mind can roam large sections of the universe of knowledge at will.

The first of the three concepts—a metalanguage of the social and natural sciences—is the intellectual counterpart of money. Just as money serves as a medium for exchanging the products of different industries or industrial specialists, this metalanguage is being designed to serve as a medium for exchanging the products—ideas—of different sciences or scientific specialists. By unifying, and thereby simplifying, the sciences, both pure research and curriculum research can more nearly be "optimized."

The second concept is the intellectual counterpart of a generalized conception of profits (in the sense of net gain) which (in a competitive market) can be used to guide, motivate, and measure effectiveness in the use of resources. This common denominator of purposes facilitates the tasks of setting priorities and integrating the varied activities of an industrial or educational enterprise.

This general economizing model is being devised to meet several requirements of a good testing, guiding, and motivating vehicle of organization. The model must provide identifiable, measurable, and demonstrable tests or yardsticks of intellectual efficiency broadly conceived. At the same time it must provide a set of guidelines for increasing efficiency and a set of motivations leading individual persons or groups to strive to be more efficient. Furthermore, this efficiency generator (or model) along with the metalanguage counterpart of money, must provide a functional basis for interrelating all inputs and outputs of any given enterprise. It must help us identify the value of inputs by relating them to the derived outputs. Finally, it must provide a basis for evaluating the results of experimental efforts and for utilizing these evaluations in designing follow-up experiments.

Although the present model relies heavily on intuitive judgments, it provides an operational basis for identifying the strategic variables and a means of organizing the relevant information once it (or an estimate of it) is available.

The third concept from economic history might be called "resourcefulness." Businessmen learn early in economic history that their success often depends more on the practical use of imagination than on the volume of economic resources immediately available. Likewise, in interdisciplinary thinking, a person's success often depends more on how imaginatively he uses his intelligence to devise ingenious strategies than it does on how high his I.Q. is. At the Center we are using the practical principles and procedures for cultivating creativity (developed and tested by the Buffalo-based Creative Education Founda-

174

tion) in the *average* person's mind and, thereby, equipping him to participate more meaningfully in interdisciplinary thinking.[11]

Perhaps the single most important implication of these three potential "quantum leaps" in education is that together they may make possible a quantum leap toward the "democratization" of interdisciplinary science. By combining these three innovations in education we hope to be able to help many culturally deprived children and adults bypass traditional educational programs and move right into the educational opportunities of the space age. Hopefully, these developments should help people generally to become masters—rather than victims—of the technological forces shaping our world.

Important as are general systems theory and some of its suggestive applications to education, they in themselves are worth relatively little in the practical sense unless they are made available to and are utilized by the peoples throughout the world who need them. Let us now turn to this challenging dimension of the task of global survival in the decades ahead.

IV. TOWARD DEVELOPING A GLOBAL "DELIVERY SYSTEM"

Insight and a Will

Needless to say, the efforts of the SGSR Task Force on General Systems Education to design a suggestive plan and strategy for evolving a delivery system for general systems theory and education are being guided to a great extent by the writings and teachings of our distinguished guest of honor. Of particular help is the following statement penned in 1957: "Where there is *insight and a will* there might be a way [italics added]."[12]

These words have especial significance for those who want to play an active role in developing an educational program which, in turn, might be used to deliver general systems theory to the people who will blueprint and implement our efforts to prevent our society from getting trapped in "the pit of meaninglessness, suicide, and atomic fire." What will be needed will not be artificial stimulants under the guise of moral suasion to pressure people into substituting the public interest for their private interest, but genuine efforts to help them discover for themselves the fact that the public interest is their private interest.

In other words, if we are to transform the power potentials of general systems theory into the reality of a better world for tomorrow, we must, I believe, work toward getting each individual to become a self-propelled, intellectual navigator, cooperatively working with others to chart the way toward a better life for all.

Joint Council on General Systems Education

We obviously cannot expect to achieve this condition in which all individuals will be willing and able to perform in this versatile, cooperative, and creative way without some special institutional arrangements established for this purpose. With this need in mind, several members of the SGSR Education Task Force are informally in the process of developing embryos of possible blueprints for globalizing general systems education. Although it would be premature to report officially on the progress of that group, I am free to sketch the heart of one dream being considered.

That overall *mission* of the Education Task Force (in cooperation with other committees of the SGSR and perhaps other organizations) is conceived to provide simply, inexpensively, and operationally the conceptual capability of converting our fragmented complex of domestic, defense, environmental, and space problems into an *integrated pattern of global opportunities*.

The *strategy* is to restructure our educational system—or at least major components thereof—in terms of general systems processes, so as to enable it to provide an operational basis for developing the global perspective, the administrative ingenuity, responsiveness, and coordination required for peaceful, wholesome living in increasingly dynamic, complex, and interdependent democracies.

The *procedure* for translating this mission and strategy into action is organized around the establishment of a national coordinating agency for developing and coordinating a national network of centers and institutes for general systems education. In the advanced planning and tentative planning of the SGSR Task Force on General Systems Education we are using the label "Joint Council on General Systems Education," because the organizational structure might be patterned after the Joint Council on Economic Education (in the U.S.A.). This is a private agency designed to coordinate an informal network of semiautonomous, voluntary, nonprofit, nonpartisan, regional, and local centers, with each center typically being sponsored and financed by a local community council. Each local council is made up of volunteers (persons and agencies) in the community representing all relevant interest groups.

Prerequisite to the development of a joint council of centers will be the development of a prototype collegiate center or institute which would not only serve as the main catalyst in developing the needed experimental research and educational programs, but also as the agency to coordinate the institutional machinery needed to effectuate the early stages of the plan.

Given a joint council on general systems education at the national level, the next two steps would be to establish the national network of centers and councils, and a global version of the joint council. If successful, the stage should then be set for interdisciplinary dialogues essential to achieving grass-roots involvement in global, participatory democracy.

I hope this interpretation of some of the potentials of general systems education will be recorded as a living testimonial to the intellectual, educational, and human power of the general systems forces which Ludwig von Bertalanffy is setting into motion. If somehow we human beings can make it to a genuine global dialogue, I believe we might have a good chance to be saved from the "pit of meaninglessness, suicide, and atomic fire."

APPENDIX

I. POSSIBLE DEGREE PROGRAMS IN GENERAL SYSTEMS STUDIES

Name of Program: The General Ecology of Knowledge (GEK)
Degree Programs:
1. Bachelor of Science in the Ecology of Knowledge (B.S.)
2. Master of Science in the Ecology of Knowledge (M.S.)
3. Doctor of Education in the Ecology of Knowledge (Ed.D.)
4. Doctor of Philosophy in the Ecology of Knowledge (Ph.D.)

Prospective *Areas for Major Concentrations* in Appropriate Programs:
1. Systems Education Studies (B.S., M.S., Ed.D., Ph.D.)
2. Basic Integrated Studies (B.S., M.S., Ed.D., Ph.D.)
3. Design Heuristics (B.S., M.S., Ed.D., Ph.D.)
4. Meta-Policy Studies (B.S., M.S., Ed.D., Ph.D.)
5. Meta-Policy Sciences (M.S., Ph.D.)

Options within Each Major Concentration:
1. Basic Research in General Ecology of Knowledge (G.E.K.)
2. Applied Ecology of Knowledge (A.E.K.)

II. CONTENT OF PROPOSED CURRICULUM CONCENTRATIONS

For each of the five concentrations listed, a few suggested topics are listed below for seminars, workshops, and institutes.

In developing the actual curriculum for any of the five areas, an effort would be made to incorporate an opportunity for each individual student to have meaningful encounters with each of the topics listed. Sometimes these encounters would be incorporated into some "traditional" subject content course. At other times a new seminar or workshop would be required. It might focus on one or more of the topics shown and would certainly overlap with other topics listed within this program and in the other four programs as well.

178

General Ecology of Knowledge in Curriculums of the Future

A. *Systems Education Studies*
   1. Systems Approaches
   2. Heuristic Problem Solving
   3. Flow-Chart Diagramming
   4. Simulation
   5. Information Technology
   6. Systems Management
   7. Game Theory
   8. Planning, Programming, Budgeting Systems
   9. PERT and Critical Path Techniques
   10. Forecasting Methods
   11. Accountability Engineering
   12. Operations Research
   13. Probing Skills
   14. Predictive Simulation

B. *Basic Integrated Studies*
   1. General Cybernetics (cybernetic process oriented around any appropriate field such as sociology, biology, etc.)
   2. Assembly of the Sciences
   3. Assembly of the Humanities
   4. Eco-Cybernetics
   5. The Control Process (as seen in traditional literature of sciences and humanities)
   6. Introduction to Meta-Tools and Meta-Skills
   7. Construction of Metalanguages
   8. General Ecology of Knowledge

C. *Global Design Heuristics*
   (This series of seminars and workshops will be designed to provide preliminary, intuitive, exploratory exercises in designing the desired society of the 21st Century and in designing alternative strategies for "reaching" these societies. The first time each course is offered, enrolled students and instructors may collectively design the course for the next "generation" of students.)
   1. Designing the Future (heuristically)
   2. Resourcefulness in Design
   3. Heuristic Forecasting
   4. Utopian Models
   5. Morphological Creativity
   6. Synergistics
   7. Synectics
   8. Designing World Games

9. Societal Engineering
10. Bionics
11. Universal Design Heuristics (applicable to poetry, music, drama, painting, biology, etc., as well as to engineering)
12. Advanced Heuristics
13. Pattern Recognition

D. *Heuristic Meta-Policy Studies*
    1. The Anatomy of Meta-Policy
    2. The Heuristics of Meta-Policy Studies
    3. Heuristic Cross-Impact Analyses
    4. Forecasting
    5. The Delphic Method
    6. The Anatomy of Change
    7. Policy Planning Methods
    8. The Psychology of Mass Movements
    9. Systems of Signals and Incentives
    10. Heuristic Simulation
    11. Designing Games for Global Grass-Roots Involvement
    12. The Process of Strategy Development
    13. The Language of Opportunity
    14. Stochastic Model Building
    15. The Dynamics of Feedback and Reinforcement

E. *Meta-Policy Sciences*
    1. Scope and Method of Meta-Policy Sciences
    2. Stochastic Model Building
    3. Behavioral Science and General Systems Perspectives
    4. Designing and Evaluating Alternative Futures
    5. Cross-Impact Analysis
    6. Setting Priorities
    7. Meta-Models and Metalanguages
    8. Normative Decision Theory
    9. Comprehensive Planning
    10. Welfare Economics
    11. Probability Theory
    12. International Games of Strategy
    13. Integrative Workshops

Response

LUDWIG VON BERTALANFFY

feel unable to do justice to the individual contributions to the present volume and the wealth of ideas offered. What is obvious is the breadth of outlook and the truly "interdisciplinary" nature of this symposium. Thus Anatol Rapoport raises fundamental questions of human knowledge, such as those of prediction and understanding, the necessity and limitations of analogical thinking, and the ultimately metaphorical character of all our thinking. This leads to the question concerning the nature of the discipline known as "general systems theory," discussed by Anatol Rapoport, Lionel J. Livesey, and others, and of the principal ways of creating such a theory in mathematical terms, exposed by Robert Rosen. The value of this general framework must be proven by its application to concrete problems, its fertility by answering old questions and posing new ones. Otherwise it remains an intellectual artifice, delightful to the theorist as the fugal development of a theme is to the musician, but without influence on science and life in general.

Thus we are especially indebted to Ekkehard Zerbst for having presented the relevant ideas with reference to their impact on physiology and related fields, through a review both of current investigations and of his own work. H. H. Pattee called our attention to unresolved problems of "anamorphosis," or the evolution of complexity. A rather different field of application is in psychology and psychiatry. Nicholas Rizzo reports on the development of systems ideas within the currents of contemporary psychology, and William Gray introduces another important aspect: general systems theory as an essentially "humanistic" endeavor. Considering the much-deplored but nevertheless progressive dehumanization of the individual in the practice of medicine, in psychiatry, and in the behavioral engineering of modern man, this humanistic emphasis can hardly be overestimated.

In a different way, Lee Thayer emphasizes the "human" aspects in communication, compared to the technological aspects of information theory, "the brain as a computer," and the like. General systems theory has, of course, a large impact on the social sciences. Kenneth Boulding, a co-founder of the movement, discusses this impact (or rather, its limitations in the conventional, "encapsulated" discipline) as it appears in his own specialty, economics. Similarly, there are close connections with management science and related fields. L. J. Livesey, a highly placed administrator of the giant educational "system" which is the State University of New York, expresses both his acceptance of the general principles and his remaining doubts and questions. Problems of education, explicitly or implicitly, ran through the series of contributions; Jere W. Clark formulates some of these in terms of curricular questions.

Looking over the vast literature now in existence on general systems theory and related topics, I do not know of another work which presents a comparably comprehensive overview.

One notices the manifold "crossing over" in contributions and investigations which—it should be remembered—originated completely independently of each other. What, for example, Rapoport, speaking as a philosophical mathematician, has to say about systems is discussed by Rosen as dynamical system models. The very same model (and equations) underlies Zerbst's survey of the physiology of the organism as "open system" and his analysis, by simulation, of problems in the physiology of the neural and sensory systems. The same is again basic, according to Boulding, to the models of quantitative economics or econometrics. The reader will do well to study these parallels in detail, and will doubtless find inspiration for work in his own specialty.

What has just been said is, of course, a paraphrase of the

well-known principles of general systems theory. It is a confirmation of what Ervin Laszlo stated in his introduction. They bear repetition, however, because this gives the present meeting what I would like to call its "symbolic meaning."

Our time is that of extreme specialization, which has led on the one hand to the resplendent achievements of science and technology, and on the other, to well-known dangers, crises, and impending catastrophes. This volume proves that it is still possible to bring together a remarkable group of people —men who are practitioners of high caliber in their fields, but nevertheless have the wide perspectives which our times obviously demand.

But the "symbolic meaning" of this symposium goes even further. We are living in a time of dissent, upheaval, revolutions and struggle, frequently aimed at mutual destruction. From the present book a different picture emerges. I have to include here many co-workers and friends who are not present, whom I do not know in person, and perhaps am not even aware of. They come not only from different disciplines, but also profess different philosophies and ideologies. They range from "tough-minded" scientists to humanistic psychologists, psychiatrists and philosophers; they include positivists, existentialists, Catholic priests, Marxists, and others. What is remarkable is the fact that, even in our time of clash of opinions and dissent, there are certain ideas and principles ("paradigms" if you like), that can be accepted by representatives of greatly divergent or even opposing convictions—and (and this is of particular importance) can be accepted without making concessions or abandoning one's persuasions. This "unity in diversity" or "coincidence of opposites" (to quote my favorite mystic, Nicholas of Cusa) is a fact far exceeding in importance the honor conferred on one individual. I am proud of having such friends, and I count them perhaps the highest gift given to me in my life.

Our symposium, fortunately, was not a laudation of my work, although this has been aptly reviewed by Laszlo and others. It has shown how many problems are left, how many perspectives are still unexplored, and how much there is to do for younger and coming generations. What was discussed was not a closed book, or "closed system," but an eminently open one—open to be elaborated by many men whose proficiency and technical knowledge exceeds my own. I am happy that this is the case; for in contrast to Professor Skinner (to whom reference was made several times) I did not find ultimate truth or "nothing-but" solutions, and never aspired toward an "ineradicable aura of quasi-religious dogma" or a secular *"extra ecclesiam nulla salus."*[1] Rather, whatever I may have been able to contribute, leaves plenty for others to do better—and I feel happy that a number of excellent men appear to be willing to do so.

Such an attitude, perhaps, partly answers Livesey's question: What is *really* "general systems theory"? This, as he notes, is partly a question of labels. I do not mind whether you call the trend in question "system theory," "the systems approach" —or "crêpe suzette." I would compare such catholic usage with terms like "theory of evolution" which covers about everything from fossil digging, classical morphology, molecular genetics, to sophisticated mathematics, without this having impeded the tremendous impact of the theory of evolution in biological science and our world view in general. Similarly, "general systems theory" offers essentially a new "paradigm,'" which is being elaborated, mathematically as far as that is possible, otherwise in verbal formulation (for mathematics is a language adapted to certain but by no means all aspects of reality) and which can be applied to a "new" class of phenomena. Concurrently the theory provides, "meta-scientifically," a new world outlook or philosophy. No *hubris* is involved; this is one of the "perspectives" human beings—with their endowments and bondages, biological, historical, linguistic, etc.—are able to form; presumably richer and better than the paradigms of the mechanistic conception or of robot psychology, but by no means a "nothing but" or "cure-all" for the ills of the individual and society.

There are a few ideas I would like to offer in the wake of Boulding's contribution. One of his key notions is that of "communication." I happen to be a student (and somewhat of a connoisseur) of postal history and the origins of mail service in early Renaissance Italy. This is a tiny and not well-known item within the wide context raised by Boulding, but it led me to some—not absolutely original—considerations.

"Centralization produces failure to optimize because of the breakdown of the communication network." A clear example is the breakdown of empires. No wonder the Spanish empire didn't do so well when it took months for a communication from the Americas to reach Seville; or Napoleon's empire when at least two weeks were required for the postal courier from the Russian steppes to Paris. (As a matter of fact, an abortive uprising took place quickly in Paris while Napoleon was in Russia.)

Modern technology has conquered the geographical impediments of communication by telegraphy, radio, and jet planes. This is obvious; but another "breakdown" is less so. It is implicit in complex hierarchical organization. All civil servants know of "channels"; they know how long it takes for, say, a request to go through until it reaches the "authorities," which, likely as not, turn out to be an ill-programmed and unfeeling computer. One reason for the failure in communication is not geographical distance, but the intricacies (and "noise") in the communication network—channels, bureaucracy, the shifting and avoiding of responsibility, or whatever you call it.

Another interesting point is Boulding's finding concerning a lack of influence of the general systems approach on economics (even though the models and equations of econometrics are typical system descriptions, identical with or transferable to descriptions of molecular, biological, ecological, etc., systems). I submit that the encapsulation of economics and its unwillingness to accept the systems approach is close to the root of the present economic crisis.

To the layman, it would appear that there are three main

187

stages in American economics—this admittedly being a gross and perhaps naïve oversimplification. The basis, of course, is free enterprise and competition—in unscientific terms, selling wares (including labor) as the market will bear and in the hope that the "invisible hand" would lead to optimal results. This worked fine until it became unbearable to workers exploited by entrepreneurs, and impracticable for the entrepreneurs who needed the government to regulate the market, international trade, worker-employer relations, and so forth. Since the principles of Adam Smith were fine for his legendary needle factory, but less so for the complex industry and commerce of modern times, Keynesian economics was given a brief try in the form of government interference according to well-known principles, limited to economics, public works, control of business cycles, government spending, etc. This did not work either, because of capitalist pressures, the dollar crisis, the deficit arising from the Vietnam war and policing the world, and so forth.

The President, no doubt, is advised by the best economists available. But are the recently introduced measures not an archetypical example of what we system theorists are wont to call "linear causal" thinking and neglect of the "system" of interactions, national and international? The refusal to reevaluate the dollar, the wage freeze, the 10 percent surcharge on imports, etc., are economic measures irrespective of connections and interactions within many social systems of which economics and the value of the dollar is only one and probably not an isolable component; it ignores the Vietnam war as an important cause of the dollar crisis, the concerns of the unions; abandons reform of health schemes; disregards political implications in the alienation of German and Japanese allies; and quasi-forces the European Common Market into an antagonistic economic bloc. Perhaps it is time that "the economic profession" should revise its stance to "view general systems with such a massive indifference."

The hour is late, not only in this day, but also in the "system" of Western civilization which has ranged from Gothic cathedrals to skyscrapers as temples of Mammon, from Thomistic theology to subatomic physics and molecular biology. It is, I believe, one task in the theory of systems to look at what—from our human standpoint—are the systems of highest order: the great civilizations.[2] This is an enormous problem into which we cannot possibly enter here. But permit me a few concluding remarks.

I am fully aware of the atrocities, injustices, and follies of our civilization. As Sorokin once calculated, our century has the heartbreaking distinction of being the most sanguinary in five thousand years of recorded history. If our civilization is doomed, so let it be. It was the fate of previous civilizations, in part nobler than ours. But let us go down in dignity.

I refuse cowardly to surrender the values that our civilization—and no other—has created. I refuse masochistically to gloat about the deficiencies of this civilization, as is presently fashionable, to admire the supposedly ideal society of the Amazon Indians, to declare, with Malcolm X, blacks as the "chosen people" of the future, to advertise a counterculture whose Third Consciousness, according to Mr. Reich, has no more to offer than rock music, drugs, fancy hairstyles and dress, promiscuous sex, and communes which are parasitic on the commercial society they detest.

Counterculture is essentially anticulture. But the Noble Savage *à la* Rousseau has never existed, neither in prehistory nor in primitive societies, and will not come into being in a future counterculture that gets rid of the present civilization with its all-too-obvious ills. We cannot shed our historical heritage—with all its goods and wrongs—without returning to the ape man of 200,000 years ago.

The same may be expressed in somewhat different terms. The basic complaint against our age is the dehumanization of man in our mechanized, industrial, and commercial society, a robot-

189

ization which makes the individual into an ever smaller wheel in Lewis Mumford's "megamachine," and control of his behavior possible by innumerable persuaders, hidden and not-so-hidden. We are told that the two presently "fastest-growing movements in psychology" are Skinnerian behaviorists on the one hand, and "humanists of group encounter" on the other.[3]

We have heard, from Drs. Livesey, Gray, Rizzo, and others, that our group of systems thinkers has little sympathy with the robot model in theory and in behavioral engineering, advertised by Skinner as the only way to salvage humanity; even though Skinner is probably right in so characterizing present trends. The opposite trend, often claiming the predicate "humanistic psychology," finds its realization in encounter groups, mind-expanding drugs, nude marathons, the human potential movement, and the like. Our psychiatric friends will agree that these measures, in part and under careful scrutiny, comprise quite legitimate and even orthodox approaches in group therapy. Besides, a totally utilitarian and routinized society, which has lost such ancient arts as flirt and conversation, may well enjoy new ways of personal and "human" contact. But there is a strange paradox involved.

Both these "movements"—antithetical as they are in all other respects—concur in one important aspect, namely, in the "zoomorphic" conception of human nature and the devaluation of the individual. Such is, of course, the essence of behavioral conditioning which, with Skinner, takes the white rat as the human "paradigm" and refuses the "myths" of "Freedom and Dignity." But strangely, it is the same goal in the opposite movement which, so to speak, dissolves the human individual in the "group," a sort of social amoeba where only "zoomorphic" drives and feelings are left. In both antitheses, what is specifically "human"—reason, culture, tradition—tends to be discarded. It is symptomatic of the *Zeitgeist* of mass society that contrary theories and practices tend toward the same result, one by making the human individual into a replaceable machine; the other by immersing him into the "group" as into a sort of social nirvana.

Whatever the future may bear, we have to preserve human values and dignity. Contrary to the recent best seller, these are neither mystical nor obsolete and unscientific superstitions, to be managed and replaced by "scientific" techniques. They are, quite simply, what is specific to man and human culture.

Notes

INTRODUCTION

1. Full bibliographical details on books by Ludwig von Bertalanffy may be found on pages 205–6.

2. Ervin Laszlo, *Introduction to Systems Philosophy: Toward a New Paradigm of Contemporary Thought* (New York, London, and Paris: Gordon and Breach, 1972).

3. Thomas S. Kuhn, *The Structure of Scientific Revolutions* (2d ed.; Chicago: University of Chicago Press, 1970). Postscript.

4. See pp. 53–4, below.

5. Ludwig von Bertalanffy, "The History and Status of General System Theory," in *Trends in General Systems Theory*, George J. Klir, ed., New York: John Wiley, 1972.

6. See the articles in this volume. Further titles are included in the Basic Bibliography, pp. 203–4.

2. THE EVOLUTION OF SELF-SIMPLIFYING SYSTEMS

1. L. von Bertalanffy, "Chance or Law," in *Beyond Reductionism* (London: Hutchinson and Co. Pub. Ltd., 1969), p. 1.

2. H. H. Pattee, "On the Origin of Macromolecular Sequences," *Biophysical J.*, 1 (1961), 683–710.

3. J. von Neumann, "The General and Logical Theory of Automata," in *The World of Mathematics*, J. R. Newman, ed., vol 4, p. 2070 (New York: Simon and Schuster, 1956). See also *The Theory of Self-reproducing Automata*, A. W. Bunks, ed., Fifth lecture (Urbana: University of Illinois Press).

4. E. P. Wigner, "On the Impossibility of a Quantum Mechanical Self-reproducing Unit," in *The Logic of Personal Knowledge* (London: Routledge and Kegan Paul, 1961), p. 231.

5. J. R. Platt, "Properties of Large Molecules that Go Beyond the Properties of their Chemical Subgroups," *J. Theoret. Biol.*, 1 (1961), 342.

6. M. Kimura, "Natural Selection as the Process of Accumulating Genetic Information in Adaptive Evolution," *Gen. Res. Camb.*, 2 (1961), 127.

7. P. S. Moorehead and M. M. Kaplan, eds., *Mathematical Challenges to the Neo-Darwinian Interpretation of Evolution* (Philadelphia: Wistar Inst. Press, 1967).

8. R. Rosen, in *Biogenesis, Evolution and Homeostasis,* A. Locker, ed., (Heidelberg and New York: Springer-Verlag, in press).

9. R. Levins, "Complex Systems," in *Towards a Theoretical Biology,* vol. 3, C. H. Waddington, ed. (Chicago: Aldine Pub. Co., 1970), p. 73.

10. H. H. Pattee, "Physical Theories of Biological Coordination." *Quart. Rev. Biophysics,* 3 (in press).

11. S. Kauffman, "Behavior of Randomly Constructed Genetic Nets," in *Towards a Theoretical Biology,* vol. 3, C. H. Waddington, ed. (Chicago: Aldine Pub. Co., 1970), p. 18.

12. R. Thom, "Topological Models in Biology," in *Towards a Theoretical Biology,* vol. 3, C. H. Waddington, ed. (Chicago: Aldine Pub. Co., 1970), 89.

13. M. Conrad and H. Pattee, "Evolution Experiments with an Artificial Ecosystem," *J. Theor. Biol.,* 28 (1970), 393.

14. H. H. Pattee, "The Recognition of Description and Function in Chemical Reaction Networks," in *Chemical Evolution and the Origin of Life,* R. Buvet and C. Ponnamperuma, eds. (North-Holland Pub. Co., 1971), p. 42.

4. THE IMPACT OF VON BERTALANFFY ON PHYSIOLOGY

1. E. S. Russel, *Science Progress* (London, 1933).

2. J. Needham, in *Nature,* 132 (1933).

3. Th. Dobzhansky, "Are Naturalists Old Fashioned?," *American Naturalist,* 100 (1966).

4. B. Commoner, "In Defense of Biology," *Science,* 133 (1961).

5. R. Dubos, "Environmental Biology," *Bioscience,* 14 (1964).

6. A. E. Needham, "The Mathematical Definition of Growth," *Unity Through Diversity,* W. Gray and N. Rizzo, eds., New York (in press); K. R. Allen, "Application of the Bertalanffy Growth Equation to Problems of Fisheries Management," *Unity Through Diversity, op. cit.;* F. Krüger, "The Energetics of Animal Growth," *Unity Through Diversity, op. cit.;* J. H. Scharf, "The Problem of the Oscillating Component of Human Growth," *Unity Through Diversity, op. cit.;* A. J. Fabens, "Properties and Fitting of the von Bertalanffy Growth Curve," *Growth,* 29 (1965).

7. R. J. H. Beverton and S. J. Holt, "On the Dynamics of Exploited Fish Populations," *Fishery Invest,* Series II, Vol. XIX (London, 1957).

8. Ludwig von Bertalanffy and Felix D. Bertalanffy, "A New Method for Cytological Diagnosis for Pulmonary Cancer," *Ann. New York Acad. Sci.* 84 (1960).

9. M. N. Meissel, "Flourescence Cytochemistry of Nucleic Acids," *Unity Through Diversity, op. cit.*

10. S. R. de Groot, *Non-equilibrium Thermodynamics* (New York: Interscience Publishers, 1962); P. Glansdorf and I. Prigogine, "On a General Evolution Criterion in Macroscopic Physics," *Physica,* 30 (1964), 351–374; A. Katchalsky, "Thermodynamics and Life," *Proceedings of the International Union of Physiological Sciences,* 8 (1971), 60–61; A. Katchalsky and R. Spangler, "Dynamics of Membrane Processes," *Quart. Rev. Biophys.,* 1

Notes

(1968), 127; I. Prigogine, *Introduction to Thermodynamics of Irreversible Processes* (New York: Interscience Publishers, 1961); "Steady States and Entropy Production," *Physica*, **31** (1965).

11. G. Oster, et al., "Network Thermodynamics," cit. in: A. Katchalsky, "Thermodynamics and Life," *Proceed. Intern. Union Physiol. Sci.*, **8** (1971), 60–61.

12. B. Chance, "The Identification and Control of Metabolic States," *Behavioral Sciences*, **15** (1970), 1–23.

13. B. Hess, "Molecular Organisation of Glycolysis," *I. Europ. Biophysics Congr. Proceed.*, **4** (1971), 447–450.

14. Ekkehard Zerbst, "Eine Methode zur Analyse und quantitativen Auswertung biologischer steady-state-Übergänge," *Experientia*, **19** (1963), 166–168. "Untersuchungen zur Veränderung energetischer Fliessgleichgewichtsübergänge bei physiologischer Anpassung." I & II. *Pflügers Arch. ges. Physiol.*, **277** (1963), 434–457; (*et al.*), "Untersuchungen zur prinzipiellen Abbildung der Sinneszellfunktionen am elektrischen Rezeptormodell." *Pflügers Arch. ges. Physiol.*, **279** (1964); *Zur Auswertung biologischer Anpassungsvorgänge mit Hilfe der Fliessgleichgewichtstheorie.* Habilitationsschrift, Freie Universität Berlin (1966); (*et al.*), "Analysis of PD–Receptor Performance. *IFAC-Internat. Symposium on Technical and Biological Problems of Control* (Yerevan USSR, 1968); (and K.-H. Dittberner), "An Analysis of the Function of Sensory Cells with the Help of Von Bertalanffy's Steady State Theory," in *Unity Through Diversity*, loc. cit.

15. Ekkehard Zerbst, "Use of an Electric Pressoreceptor Analog in Baropacing of Hypertension in Bloodpressure." *Proceed. III Internat. Congr. Biophysics*, Cambridge, Mass. (1969); "Principes et résultats de la barostimulation en circuite fermé au moyen d'un stimulateur électrique," *Ann. Cardiol. Angéol.*, **20** (1970), 105–112.

16. Walter Beier, *Biophysik*, 3rd ed. (Leipzig: Georg Thieme, 1968); F. H. Dost, *Der Blutspiegel* (Leipzig: Georg Thieme, 1953); 2nd enlarged edition *Grundlagen der Pharmakokinetik* (Leipzig: Georg Thieme, 1968); E. Malek et al., *Continuous Cultivation of Microorganisms* (Prague: Czech. Acad. Sci. 1958, 1964); A. Rescigno and G. Segre, *Drug and Tracer Kinetics* (Waltham, Mass.: Blaisdell, 1966); Robert Rosen, *Dynamical System Theory in Biology*. Vol. I (New York: Wiley, 1970).

17. R. A. Smith, III, in *General Systems*, **15** (1970), 239.

5. ECONOMICS AND GENERAL SYSTEMS

1. K. E. Boulding, "An Application of Population Analysis to the Automobile Population of the United States," *Kyklos*, **2** (1955), 109–124.

6. COMMUNICATION SYSTEMS

1. Ludwig von Bertalanffy, "The History and Status of General System Theory," in George J. Klir, ed., *Trends in General Systems Theory* (New York: Wiley, 1972).

2. As in the paper cited above; in all references to "communication" in *General System Theory: Foundations, Development, Applications* (New York: Braziller, 1968); and elsewhere.

3. As evidenced, e.g., in the M. D. Mesarović-edited volume, *Views on General Systems Theory* (New York: Wiley, 1964); F. Kenneth Berrien, *General and Social Systems* (New Brunswick: Rutgers University Press, 1968); John H. Milsum, ed., *Positive Feedback: A General Systems Approach to Positive/Negative Feedback and Mutual Causality* (London: Pergamon, 1968); and, in general, the annual issues of *General Systems* since 1956, as well as collections like Walter Buckley's *Modern Systems Research for the Behavioral Scientist* (Chicago: Aldine, 1968).

4. A notable exception here, of course, is Walter Buckley's *Sociology and Modern Systems Theory* (Englewood Cliffs, N. J.: Prentice-Hall, 1967), and some of the authors represented in the William Gray-, F. J. Duhl-, and N. D. Rizzo-edited volume, *General Systems Theory and Psychiatry* (Boston: Little, Brown, 1969), such as Jurgen Ruesch and Albert E. Scheflen.

5. E.g., "On the Definition of the Symbol," in Joseph R. Royce, ed., *Psychology and the Symbol: An Interdisciplinary Symposium* (New York: Random House, 1965); and in *Robots, Men and Minds: Psychology in the Modern World* (New York: Braziller, 1967), the latter based upon the Heinz Werner Inaugural Lectures in 1966, published by Clark University Press in 1968 under the title, *Organismic Psychology and Systems Theory*.

6. I would also argue that Von Bertalanffy didn't go as far as he perhaps should have in redefining man's symbolic behavior in general systems terms.

7. *Cf.* Anatol Rapoport, "Man, The Symbol-User," in L. Thayer, ed., *Communication: Ethical and Moral Issues* (New York: Gordon and Breach, in press).

8. E.g., Kenneth E. Boulding, "General Systems as a Point of View," in M. D. Mesarović, *op. cit.*; Anatol Rapoport, "Modern Systems Theory—An Outlook for Coping with Change," in *General Systems*, **15** (1970); Walter Buckley, *Sociology and Modern Systems Theory* (*loc. cit.*); Ervin Laszlo, *Introduction to Systems Philosophy* (New York: Gordon and Breach, 1972); Geoffrey Vickers, *Freedom in a Rocking Boat* (London: Allen Lane, 1970, esp. the appendix, "The Oddity of Historical Systems"); David Easton, *A Systems Analysis of Political Life* (New York: Wiley, 1958); W. Gray and N. Rizzo, eds., *Unity Through Diversity* (New York: Gordon and Breach, in press). One could hardly overlook here the contributions of the several authors in the earlier Grinker-edited volume, *Toward a Unified Theory of Human Behavior*, first published in 1956 (New York: Basic Books).

9. Particularly notable here, I think, and for purposes of comparison in Von Bertalanffy's own field of biology, are Walter M. Elsasser, *Atom and Organism: A New Approach to Theoretical Biology* (Princeton: Princeton University Press, 1966); Garret Hardin, *Nature and Man's Fate* (New York: Rinehart, 1959); and Theodosius Dobzhansky, *Mankind Evolving* (New Haven: Yale University Press, 1962). But *cf.* Jacques Monod, *Le Hasard et la Nécessité* (Paris: Seuil, 1970).

10. In *General System Theory* (op. cit.), pp. 24–25.

11. Or, in systems terms, science predicts only those systems which are closed, or which it has the technology (or the "convention") to close. *Cf.*, e.g., this comment by Stafford Beer: "The fact is that if a system is only relatively isolated (that is to say, *open*), it has to be absolutely isolated (that is to say, *closed*) by an artificial convention before its mode of control by the natural laws of cybernetics can be discussed." *Decision and Control* (New York: Wiley, 1966), p. 288.

12. Joseph H. Monane, *A Sociology of Human Systems* (New York: Appleton-Century-Crofts, 1967); and C. West Churchman, *The Systems Approach* (New York: Delacorte, 1968).

13. More extensive discussions of the conceptual difficulties involved are offered in the author's "On Theory-Building in Communication: I. Some Conceptual Problems," *Journal of Communication*, 13 (1963), 217–235; ". . . II. Some Persistent Obstacles," in J. Akin *et al.*, eds., *Language Behavior* (The Hague: Mouton, 1970), pp. 34–42; and ". . . IV. Some Observations and Speculations," *Systematics*, 1970, 7, pp. 307–314.

14. A full discussion of the faultiness and inadequacy of this common model of communication would be out of place here. But those who are interested may find a more complete argument in the author's "Communication: *Sine qua non* of the Behavioral Sciences," in D. L. Arm, ed., *Vistas in Science* (Albuquerque, N. M.: New Mexico University Press), 1968, pp. 48–77; and *Communication and Communication Systems* (Homewood, Ill.: Irwin), 1968.

15. See my "Communication—*Sine qua non* . . . ," *loc cit.*

16. It is most unfortunate, I believe, that we do not make this distinction between communication and intercommunication either in our everyday or in our scientific talk about communication.

17. There are a great many points of convergence between general systems theory and communication systems theory. As a matter of expedience, therefore, I will be emphasizing largely only the major points of *divergence* between the two.

18. Geoffrey Vickers, "A Classification of Systems," *General Systems*, 15, (1970), pp. 3–6; and *op. cit.*, fn. 8.

19. Burkart Holzner, *Reality Construction in Society* (Cambridge, Mass.: Schenkman, 1968), particularly Ch. IV.

20. *Cf.* Albert E. Scheflen, "Patterns," "Systems and Psychomatics," in William Gray *et al.*, eds., *op. cit.*, pp. 159–70.

21. For these pseudo- but provocative acronyms, I am indebted to Mel Thistle. See, e.g., his "Emotional Barriers to Communication," in L. Thayer, ed., *Communication—Spectrum '7: Proceedings* of the 15th Annual Conference of the National Society for the Study of Communication (published by the Society, 1968), pp. 268–78.

22. John N. Bleibtreu, for example, refers to ". . . the evolutionary role of excommunication as a stimulator of diversity," in *The Parable of the Beast* (New York: Macmillan, 1968), p. 164. *Cf.* Paul Weiss, quoted in

Walter M. Elsasser, *op. cit.*, pp. 63–64; and Ernst Mayr, *Animal Species and Evolution* (Cambridge, Mass.: Harvard University Press, 1963).

23. Some corollaries are described in the author's "Communication," in Rubin Gotesky and Ervin Laszlo, eds., *Evolution–Revolution: Patterns of Development in Nature, Society, Man, and Knowledge* (New York: Gordon and Breach, 1971).

24. Some of the moral-ethical issues involved are dealt with in the author's "Toward an Ethics of Communication," *Revista de Occidente,* 1971; and by numerous scholars and thinkers in L. Thayer, ed., *Communication: Ethical and Moral Issues* (*loc cit.*, fn. 7).

7. BERTALANFFIAN PRINCIPLES AS A BASIS
FOR HUMANISTIC PSYCHIATRY

1. William Gray, "General System Theory as a Humanistic Science." Presented at the Symposium of the Psychology Department of York University, Ontario, Canada, March 12, 1971; "Nurturing the Humanity of Man: A Metalanguage for Court Clinics." Presented at the Symposium on Conceptual and Attitudinal Pathways from Man to Mankind, sponsored by the Center for Interdisciplinary Creativity, Southern Connecticut State College, and the Society for General Systems Research, Windsor, Connecticut, April 29, 1971; "Ludwig von Bertalanffy's General System Theory as a Model for Humanistic System Science." Presented at the XIIIth International Congress of the History of Science, Section 1, Subsection "History of System Analysis," Moscow, U.S.S.R., August 18–24, 1971.

2. Walter Buckley, personal communication, August, 1971.

3. Ervin Laszlo, *System, Structure, and Experience* (New York: Gordon and Breach, 1969).

4. Ludwig von Bertalanffy, *General System Theory* (New York: Braziller, 1968), p. 192.

5. *Ibid.*, pp. 193–194.

6. Seymour L. Halleck, *The Politics of Therapy* (New York: Science House, 1971).

7. L. J. West, "Ethical Psychiatry and Bio-Social Humanism," *Am. J. Psychiat,* **126** (1969), 112–116.

8. "Correspondence of Aldous Huxley and Ludwig von Bertalanffy," in *Unity Through Diversity: A Festschrift for Ludwig von Bertalanffy,* Vol I, *op. cit.*

9. Nicholas D. Rizzo, William Gray, and Julian S. Kaiser, "A General Systems Approach to Problems in Growth and Development," in *General Systems Theory and Psychiatry,* edited by William Gray, Frederick J. Duhl, and Nicholas D. Rizzo (Boston: Little, Brown, 1969).

10. William Gray, "Protocol for a Humanitarian Review Board at the First District Court of Eastern Middlesex." September, 1971.

11. Nicholas D. Rizzo, personal communication, February, 1971.

12. Halleck, *op. cit.*

Notes

13. In 1967 Von Bertalanffy was elected as Honorary Fellow of the American Psychiatric Association, and with this his influence has continued to spread. The list of psychiatrists taking part in the development of humanistic psychiatry continues to grow and now includes, besides such pioneers as Karl Menninger, Roy R. Grinker, Sr., Silvano Arieti, James G. Miller, Jurgen Ruesch, John P. Spiegel, and those already mentioned in this paper, the names of Edgar H. Auerswald, Warren M. Brodey, Edward J. Carroll, E. Joseph Charny, Frederick J. Duhl, Leonard J. Duhl, Norris Hansell, Don D. Jackson, H. Peter Laqueur, John MacIver, Judd Marmor, Jules H. Masserman, Norman L. Paul, Eugene Pumpian-Mindlin, Howard P. Rome, Albert E. Scheflen, Montague Ullman, and Raymond W. Waggoner, Sr. Additionally I would mention Helen E. Durkin, who is active in the group psychotherapy movement.

8. THE SIGNIFICANCE OF VON BERTALANFFY FOR PSYCHOLOGY

1. *Encyclopedia Britannica,* **10** (1968), 370–371.

2. See *Robots, Men and Minds* (New York: George Braziller, 1967); *General System Theory* (New York: George Braziller, 1968); and "General System Theory and Psychiatry," *American Handbook of Psychiatry,* ed. S. Arieti, vol. III (New York: Basic Books, 1966).

3. R. I. Evans, *B. F. Skinner, The Man and His Ideas* (New York: Dutton, 1968).

4. N. D. Rizzo, H. M. Fox, S. Gifford, B. J. Murawski, and E. N. Kudarauskas, "Some Methods of Observing Humans Under Stress," *Psychiatric Research Reports,* **7** (April, 1957).

5. N. D. Rizzo, W. Gray, and J. S. Kaiser, "A General Systems Approach to Problems in Growth and Development," *General Systems Theory and Psychiatry,* ed. W. Gray, F. J. Duhl and N. D. Rizzo (Boston: Little Brown, 1969); N. D. Rizzo, "The Court Clinic and Community Mental Health: General System Theory in Action," *Unity Through Diversity, op. cit.,* vol. II; N. D. Rizzo, "Il Principio Fondamentale della Psichoterapia Non-Volontaria" (Milan: International Congress of Psychotherapy, in press); N. D. Rizzo, D. H. Russell, and W. Gray, "Theoretical and Practical Aspects of Enforced Psychotherapy," American Correctional Association, Centennial Meeting, October, 1970 (in press); and N. D. Rizzo and L. von Bertalanffy, "General System Theory: Its Impact in the Health Fields," Conference on Health Research and the Systems Approach, Wayne State University, March 1, 1971.

9. NOETIC PLANNING: THE NEED TO KNOW, BUT WHAT?

1. Kenneth E. Boulding, *Beyond Economics* (Ann Arbor, Michigan: The University of Michigan Press, 1969).

2. George A. Miller, Eugene Galanter, and Karl H. Pribram, *Plans and the Structure of Behavior* (New York: Holt, Rinehart and Winston, 1960).

3. Kenneth E. Boulding, "General Systems Theory—The Skeleton of Science," *Management Science,* **2** (1956), 197–208.

4. Theodosius Dobzhansky, *Mankind Evolving* (New Haven, Conn.: Yale University Press, 1962).

5. Ludwig von Bertalanffy, in Foreword to Ervin Laszlo, *Introduction to Systems Philosophy* (New York: Gordon and Breach, 1972).

6. *Ibid.,* "System, Symbol and the Image of Man" in Iago Galdston, ed., *The Interface Between Psychiatry and Anthropology* (New York: Brunner/ Mazel, 1971).

7. Ervin Laszlo, *Introduction to Systems Philosophy, op. cit.*

8. Ludwig von Bertalanffy, "System, Symbol and the Image of Man" *op. cit.*

9. Karl W. Deutsch, "Some Notes on Research on the Role of Models in the Natural and Social Sciences," *Synthese,* **7,** (1948–49), 506–533.

10. Walter Buckley, "Society as a Complex Adaptive System," in Walter Buckley, ed., *Modern Systems Research for the Behavioral Scientist* (Chicago: Aldine Publishing Co., 1968).

11. William Irwin Thompson, *At the Edge of History* (New York: Harper and Row, 1971). [Italics added.]

12. John D. Carroll, "Noetic Authority," *Public Administration Review* (September–October, 1969), 492–500.

13. Paul A. Weiss, "Living Nature and the Knowledge Gap," *Saturday Review,* **19** (November 29, 1969).

14. Robert Jungk, "About Mankind 2000," in Sanford Anderson, ed., *Planning for Diversity and Choice* (Cambridge, Mass.: M.I.T. Press, 1968).

15. Karl Mannheim, "Freedom Under Planning," in Warren G. Bennis, Kenneth D. Benne, and Robert Chin, eds., *The Planning of Change* (New York: Holt, Rinehart, and Winston, 1966).

16. Jurgen Moltmann, *Hope and Planning* (New York: Harper and Row, 1971).

10. THE GENERAL ECOLOGY OF KNOWLEDGE IN CURRICULUMS OF THE FUTURE

1. Ludwig von Bertalanffy, "Human Values in a Changing World," in Abraham H. Maslow, ed., *New Knowledge in Human Values* (New York: Harper and Row, 1959), p. 72.

2. *Ibid.,* pp. 67–68.

3. *Ibid.,* p. 74.

4. *Ibid.,* p. 73.

5. *Ibid.,* p. 73 [italics added].

6. Several of these limitations of "the systems approach" are given in Klaus Hinst's apology for using the systems approach in his article, "Edu-

Notes

cational Technology—Its Scope and Impact . . . ," *Educational Technology* (July, 1971), p. 40.

7. Ludwig von Bertalanffy, *Robots, Men and Minds* (New York: Braziller, 1967), p. 63.

8. For a list of persons who are known to be working on some of these and other concepts with this purpose in mind (and who are members of the SGSR Education Task Force), see Jere W. Clark, "Eco-Cybernetics: The Nucleus of Space-Age Knowledge and Curriculum Patterns," in Jere W. Clark, ed., *Practical Action Programs in Education* (Highlights of the Third National Conference on General Systems Education), The Calvin K. Kazanjian Economics Foundation, 1971, Appendix B.

9. For an elaboration of the positive content of the above paragraphs of this section, see Jere W. Clark, "Systems Philosophy and the Crisis of Fragmentation in Education," an appendix to Ervin Laszlo, *Introduction to System Philosophy: Toward a New Paradigm of Contemporary Thought* (New York: Gordon and Breach, 1972).

10. Here we use a broader, more heuristic, and less rigid notion of cybernetics—which I prefer to call "paracybernetics"—than the professional cyberneticians use. In this context, we might say that paracybernetics is the systematic, behavioristic, metaphoric study of communications in physical, biological, and social organizations with emphasis on regulation and control through feedback and related processes.

11. For an elaboration of this view, see Jere W. Clark, "Creativeness—Can It Be Cultivated," *The Business Quarterly* (Spring, 1965), pp. 29–39.

12. Ludwig von Bertalanffy, "Human Values in a Changing World," in Abraham Maslow, *op. cit.*

RESPONSE

1. R. L. Rubenstein, review of B. F. Skinner, "Beyond Freedom and Dignity," *Psychology Today* (September, 1971), pp. 28ff.

2. The author—who wrote about Oswald Spengler's *Decline of the West* soon after it appeared (1924)—has recently discussed "Cultures as Systems" before the American Historical Association, Annual Meeting, December 30, 1971 (New York City).

3. T. G. Harris, "All the World's a Box," *Psychology Today* (August, 1971), p. 33.

Basic Bibliography

A Selected List of Recent Books in the Area
of General Systems Theory

W. G. Bennis, K. D. Benne, and R. Chin, eds., *The Planning of Change* (New York: Holt, Rinehart and Winston, 1962).

F. K. Berrien, *General and Social Systems* (New Brunswick: Rutgers University Press, 1968).

Ludwig von Bertalanffy, *General System Theory* (New York: Braziller, 1968).

————, *Problems of Life* (New York: Wiley, 1952).

————, *Robots, Men and Minds* (New York: Braziller, 1967).

————, and Anatol Rapoport, eds., *General Systems* (Washington: Soc. for General Systems Research, 16 vols. since 1956).

Kenneth E. Boulding, *Beyond Economics* (Ann Arbor: University of Michigan Press, 1968).

————, *The Organizational Revolution* (New York: Harper, 1953).

Walter Buckley, ed., *Modern Systems Research for the Behavioral Scientist* (Chicago: Aldine, 1968).

————, *Sociology and Modern Systems Theory* (Englewood Cliffs: Prentice-Hall, 1967).

C. West Churchman, *The Systems Approach* (New York: Delacorte, 1968).

N. J. Demerath and R. A. Peterson, eds., *System, Change and Conflict* (New York: Free Press, 1967).

David Easton, *A Systems Analysis of Political Life* (New York: Wiley, 1965).

Walter M. Elsasser, *Atom and Organism* (Princeton: Princeton University Press, 1966).

F. E. Emory, ed., *Systems Thinking* (England: Penguin Books, 1969).

H. von Foerster and G. W. Zopf, eds., *Principles of Self-Organization* (New York: Pergamon, 1962).

William Gray, F. D. Duhl, and N. D. Rizzo, eds., *General Systems Theory and Psychiatry* (Boston: Little, Brown, 1969).

William Gray and Nicholas D. Rizzo, eds., *Unity Through Diversity: A Festschrift in Honor of Ludwig von Bertalanffy*, 2 Vols (New York, London, and Paris: Gordon and Breach, in press).

Roy R. Grinker, ed., *Toward a Unified Theory of Human Behavior* (New York: Basic Books, 1956).

George J. Klir, ed., *Trends in General Systems Theory* (New York: Wiley, 1972).

K. Knorr and S. Verba, eds., *The International System* (Princeton: Princeton University Press, 1961).

Arthur Koestler and J. R. Smythies, eds., *Beyond Reductionism* (New York: Macmillan, 1969).

Ervin Laszlo, *System Structure and Experience* (New York, London, and Paris: Gordon and Breach, 1969).

———, *Introduction to Systems Philosophy* (New York, London, and Paris: Gordon and Breach, 1972).

———, *The Systems View of the World* (New York: Braziller, 1972).

Abraham H. Maslow, ed., *New Knowledge in Human Values* (New York: Harper and Row, 1966).

M. D. Mesarović, ed., *Views on General Systems Theory* (New York: Wiley, 1964).

G. A. Miller, E. Galanter, and K. H. Pribram, *Plans and the Structure of Behavior* (New York: Holt, Rinehart and Winston, 1960).

John H. Milsum, ed., *Positive Feedback: A General Systems Approach to Positive/Negative Feedback and Mutual Causality* (London: Pergamon, 1968).

Talcott Parsons, E. A. Shils, K. D. Naegele, and T. R. Pitts, eds., *Theories of Society* (New York: Free Press, 1961).

Robert Rosen, *Optimality Principles in Biology* (New York: Plenum Press, 1967).

———, *Dynamical System Theory in Biology,* Vol. I (New York: Wiley, 1970).

James N. Rosenau, ed., *Linkage Politics* (New York: Free Press, 1969).

Herbert A. Simon, *The Sciences of the Artificial* (Cambridge, Mass.: M.I.T. Press, 1969).

P. A. Sorokin, *Sociological Theories of Today* (New York: Harper, 1966).

Lee Thayer, ed., *Communication: The Ethical and Moral Issues* (New York, London, and Paris: Gordon and Breach, in press).

Geoffrey Vickers, *Freedom in a Rocking Boat* (London: Allen Lane, 1970).

C. H. Waddington, ed., *Towards a Theoretical Biology* (Chicago: Aldine, 1970).

Paul A. Weiss, *Dynamics of Development* (New York: Academic Press, 1968).

———, ed., *Hierarchically Organized Systems in Theory and Practice* (New York: Hafner, 1971).

Lancelot L. Whyte, A. G. Wilson, and D. Wilson, eds., *Hierarchical Structures* (New York: Elsevier, 1969).

Books by
Ludwig von Bertalanffy

Modern Theories of Development. Transl. by J. H. Woodger. Oxford: Oxford University Press, 1933. Torchbook Edition: New York: Harper, 1962.

German: *Kritische Theorie der Formbildung.* Berlin: Gebrüder Borntraeger, 1928.

Spanish: *Teoria del Desarrollo Biologico.* Transl. by M. Biraben. Buenos Aires: Universidad de La Plata, 2 volumes, 1934.

Nikolaus von Kues. München: Georg Müller, 1928.

Lebenswissenschaft und Bildung. Erfurt: Kurt Stenger, 1930.

Theoretische Biologie. Erster Band: *Allgemeine Theorie, Physiko-Chemie, Aufbau und Entwicklung des Organismus.* Berlin: Gebrüder Borntraeger, 1932.

Zweiter Band: *Stoffwechsel, Wachstum.* Berlin: Gebrüder Borntraeger, 1942. Second edition enlarged: Bern: A. Francke AG., 1951.

Das Gefüge des Lebens. Leipzig: Teubner, 1937.

Vom Molekül zur Organismenwelt. Grundfragen der modernen Biologie. 2nd edition. Potsdam: Akademische Verlagsgesellschaft Athenaion, 1949.

Biologie und Medizin. Wien: Springer Verlag, 1946.

Problems of Life. An Evaluation of Modern Biological Thought. New York: J. Wiley & Sons, and London: Watts & Co., 1952.

German: *Das Biologische Weltbild.* Bern: A. Francke AG., 1949.

Torchbook Edition: New York: Harper, 1961.

Japanese: Tokyo: Misuzu Shobo Co., 1954.

French: *Les Problèmes de la Vie.* Transl. by M. Deutsch. Paris: Gallimard, 1960.

Spanish: *Concepción Biológica del Cosmos.* Transl. by Faustino Cordón. Santiago: Ediciones de la Universidad de Chile, 1963.

Dutch: *Een Biologische Wereldbeeld. Het Verschijnsel Leven in Natuur en Wetenschap.* Transl. by P. G. E. Schücking Kool. Utrecht: Holland, Bijleveld, 1965.

Auf den Pfaden des Lebens. Ein biologisches Skizzenbuch. Wien: Universum Verlag, 1951.

Biophysik des Fliessgleichgewichts. Transl. by W. H. Westphal. Braunschweig: Vieweg, 1953.

Robots, Men and Minds. Psychology in the Modern World. New York: Braziller, 1967.

German: . . . *aber vom Menschen wissen wir nichts.* (Enlarged edition.) Düsseldorf: Econ Verlag, 1970.

Italian: *Il Sistema Uomo. La Psicologia nel Mondo Moderno.* Transl. by L. Occhetto Baruffi. Milano: Istituto Librario Internazionale, 1971.

Japanese: *Human Being and Robot. Psychology in the Modern World.* Transl. by K. Nagano. Tokyo: Misuzu Shobo, 1971.

Spanish: *Robots, Hombres y Mentes.* Transl. by F. Calleja. Madrid: Guadarrama, 1971.

Czech: Transl. by J. Kamarýt. Prague: Svoboda, 1972.

Organismic Psychology and Systems Theory. Worcester (Mass.): Clark University Press, 1968.

General System Theory. Foundations, Development, Applications. New York: Braziller, 1968. Enlarged edition: London: Allen Lane, The Penguin Press, 1971, and New York: Braziller, 1972.

Italian: *Teoria Generale dei Sistemi.* Transl. by E. Bellone. Milano: Istituto Librario Internazionale, 1972.

Swedish: Stockholm: Wahlström and Widstrand, 1972.

Japanese: Transl. by K. Nagano. Tokyo: Misuzu Shobo, 1972.

Spanish: Madrid: Guadarrama, 1972.

French: Paris: Dunod, 1972.

German: Braunschweig: Vieweg, 1973 (in press).

Editor of:

Handbuch der Biologie (with Fritz Gessner). Frankfurt am Main: Akademische Verlagsgesellschaft Athenaion, 14 vols. (with *c.*60 co-workers) since 1942, 234 issues to-date.

General Systems (with Anatol Rapoport). Washington, D.C.: Society for General Systems Research, 16 volumes since 1956.

EDITOR'S NOTE: A complete bibliography of Von Bertalanffy's scientific writings, including papers and articles (about 280 items), is found in Wm. Gray and N. D. Rizzo, eds., *Unity Through Diversity* (see *Basic Bibliography*).

Notes on Contributors

ERVIN LASZLO is Professor of Philosophy at the State University of New York's College of Arts and Science at Geneseo, the Editor of the International Library of Systems Theory and Philosophy, and the author of *Introduction to Systems Philosophy*.

ANATOL RAPOPORT is Professor of Mathematical Psychology at the University of Toronto and Consultant to the Mental Health Research Institute at the University of Michigan. He is the Editor of the *General Systems Yearbooks* (with Ludwig von Bertalanffy) and the author of numerous books and articles.

HOWARD H. PATTEE is Professor at the W. W. Hansen Laboratories of Physics at Stanford University and has been Visiting Professor at the Center for Theoretical Biology at the State University of New York at Buffalo for 1971–72. He is editor of *Hierarchy Theory: The Challenge of Complex Systems*, shortly to be published in the International Library of Systems Theory and Philosophy.

ROBERT ROSEN is Professor at the Center for Theoretical Biology at the State University of New York at Buffalo and has been Visiting Professor at the University of California at Santa Barbara for 1971–72. He is the author of *Optimality Principles in Biology* and other books and articles.

EKKEHARD ZERBST is Professor of Physiology and Director of the Physiological Institute at the Free University of Berlin and the author of numerous works in biology and physiology.

KENNETH E. BOULDING is Professor of Economics at the Institute of Behavioral Science at the University of Colorado. His numerous works in economics and general systems theory include *The Organizational Revolution* and *Beyond Economics*.

LEE THAYER is Gallup Professor of Communication and Director of the Center for the Advanced Study of Communication in the University of Iowa, the author of numerous works in communication theory, and the Editor of the forthcoming journal *Communication*.

207

WILLIAM GRAY is a practicing psychiatrist and the Editor of *General Systems Theory and Psychiatry* as well as of the forthcoming two-volume Festschrift for Ludwig von Bertalanffy, *Unity Through Diversity* (both with Nicholas D. Rizzo).

NICHOLAS D. RIZZO is likewise a practicing psychiatrist and the Editor (with William Gray) of *General Systems Theory and Psychiatry* and *Unity Through Diversity.*

LIONEL J. LIVESEY, who died shortly after delivering the paper here published, was Vice Chancellor for Long-Range Planning of the State University of New York.

JERE W. CLARK is Director of the Center for Interdisciplinary Creativity at Southern Connecticut State College, and Chairman of the Task Force on Education of the Society for General Systems Research.

Index

Index

Index

Self-reflexivity, 105, 107
Self-simplification, 32–41
Shakespeare, W., 162
Shands, H., 117
Shannon, C., 96
Simon, H., 83
Simplicity, 14–30, 36, 37
Simplification, 61, 173
Skinner, B.F., 190
Smith, R.A. III, 75
Social science, 19–22, 79
Sociology, 79, 85, 98
Society for General Systems Research, 79, 165, 176, 177
Sorokin, P., 189
Spangler, R., 71
Specialization, 34, 185
Spontaneity, 130
Stability, 41, 81
State variables, 50, 51, 57
Stratification, 60
Structural studies, 61
Structuralism, 63
Structure, 29, 33, 34, 39, 45
Survival, 16
Symbols, Symbolism, 97, 105, 127, 131, 165
System analogies, 55–59
System criteria, 118–120
System properties, 53
Systems, classes of, 28, 29
Systems education, 159, 166, 171
Systems movement, 4, 5, 10
Systems philosophy, 7, 9, 10
Systems research, 7
Systems science, 8, 9
Systems, self-reproducing, 35
Systems technology, 7, 9
Systems theory, mathematical, 24

Task Force on Education (SGSR), 176, 177

Technologies of communication, 109
Teleology, 105, 138, 141
Telesitic, 105
Thayer, L., vi, 9, 93–121, 184
Thirteenth World Congress of History of Science, 8
Thom, R., 40
Thompson, W.I., 157
Three-body problem, 5, 53, 54

Understanding, 17–19, 21, 102
Unifying concepts, 170, 173
Unity in diversity, 185
Unity of nature, 6

Validation, 112
Values, 10, 133, 165, 167, 171, 189
Variational principles, 56
"Verstehen," 23
Vickers, G., 98, 111
Vico, G., 97
Vitalism, 141

Waddington, C.H., 10
Walras, L., 80, 81
Weaver, W., 5, 96
Weiss, P.A., 159
Werner, H., 138
Wertheimer, M., 138
West, L.J., 138
Wiener, N., 10, 96
Wigner, E., 36, 61
Wittgenstein, L. von, 45
Word, 27
World picture, 3, 4
Wright, S., 83
Wundt, W., 138, 141

"Zeitgeist," 190
Zerbst, E., vi, 3, 69–75, 183, 184
Zoomorphism, 131, 140, 143, 190